UNDERSTANDING EVOLUTION

Sixth Edition

UNDERSTANDING EVOLUTION

E. Peter Volpe
Mercer University
School of Medicine

Peter A. Rosenbaum
State University of New York at Oswego

Boston Burr Ridge, IL Dubuque, IA Madison, WI New York San Francisco St. Louis
Bangkok Bogotá Caracas Lisbon London Madrid
Mexico City Milan New Delhi Seoul Singapore Sydney Taipei Toronto

McGraw-Hill Higher Education

A Division of The McGraw-Hill Companies

UNDERSTANDING EVOLUTION, SIXTH EDITION

3 4 5 6 7 8 9 0 QPF/QPF 0 9 8 7 6 5 4 3 2 1 0

ISBN 0–697–05137–4

Vice president and editorial director: *Kevin T. Kane*
Publisher: *Michael D. Lange*
Senior sponsoring editor: *Margaret J. Kemp*
Developmental editor: *Donna Nemmers*
Marketing manager: *Michelle Watnick*
Project manager: *Mary Lee Harms*
Production supervisor: *Sandy Ludovissy*
Designer: *Rick Noel*
Senior photo research coordinator: *Lori Hancock*
Compositor: *Precision Graphics*
Typeface: *10/12 Times Roman*
Printer: *Quebecor Printing Book Group/Fairfield, PA*

Cover designer: *Nicole Dean*
Cover image: © *Tone Stone Images*
Photo research: *Mary Reeg Photo Research*

Chameleons are lizards found mostly in the Eastern Hemisphere. The chameleon on the front cover is a jewel chameleon, found on the island of Madagascar in the Indian Ocean, off the southeast coast of Africa.

Chameleons are some of the oldest reptiles on earth, dating back 150 million years to the dinosaur age. Over 100 chameleon species exist, the majority of which are endemic to the island of Madagascar. Most species are characterized by bulging eyes, colorful skin, a prehensile tail, and Y-shaped feet. A chameleon's size can vary from $3/4$ of an inch to $2 1/2$ feet in length, although most are between 3 and 8 inches long. These lizards are known for their ability to change colors, and can blend in with almost any background. Their colors are usually green, brown, and gray, but chameleons can also turn shades of yellow, orange, black, white, and even blue. Color changes are usually related to light, temperature, or emotions, but they are most colorful at the beginning of breeding season.

Chameleons are diurnal creatures, and most of their time is spent basking in the sun, since they are cold-blooded. Almost all chameleons are tree-dwellers, since their hands and feet are well adapted to grasping branches and climbing. Chameleons are territorial and prefer to live alone. They wait in trees for prey to come to them, and catch the insects with their long, sticky tongues. Water is licked from leaves, since chameleons do not have the ability to lap water from ponds.

Library of Congress Cataloging-in-Publication Data

Volpe, E. Peter (Erminio Peter)
    Understanding evolution / E. Peter Volpe, Peter A. Rosenbaum. —
6th ed.
        p.   cm.
    Includes bibliographical references and index.
    ISBN 0–697–05137–4
    1. Evolution (Biology) I. Rosenbaum, Peter Andrew, 1952– .
    II. Title.
    QH366.2.V64   2000
    576.8—dc21   98–55197
                                                        CIP

www.mhhe.com

To those readers who are
seeking promising answers
rather than final ones

BRIEF CONTENTS

CONTENTS

CHAPTER **21**

CULTURAL EVOLUTION   *231*

EPILOGUE   *240*

"Scientific Creationism" is Not Science   *240*

PREFACE

This new edition of *Understanding Evolution*, like the earlier ones, is addressed to the students for whom it is written. We have endeavored to present a simple, concise account of the scope and significance of evolution for those college students who have had no previous experience with the subject. As an introduction to the principles of evolution, this nonencyclopedic book is ideally suited for the liberal education of both the nonscientist and scientist. This small book can stand in good stead in an introductory-level evolution course or as supplement to a basic biology or basic anthropology course. It has been rewritten and expanded to transmit the wealth of new ideas in evolution fostered by the molecular revolution in biology.

The publication of Charles Darwin's *The Origin of Species* in 1859 is one of the great landmarks in the history of science. Darwin brought about one of the most remarkable and far-reaching revolutions in human thought. Before Darwin, we presumed that humans occupied an exalted position in the world. Darwin removed us from the center stage of the earth and made us part of the fabric of nature. Although Darwin's monumental book was greeted at first with a storm of criticism, his findings are now universally acknowledged.

Darwin's thesis has been fortified by an ever-expanding knowledge of the gene. Until the past few decades, the chemical nature of the gene remained enigmatic. To many biologists, the gene was just an intellectual device to organize data.

One of the finest triumphs of modern science has been the elucidation of the chemical makeup of the gene. The fundamental chemical component of the gene is the remarkable molecule, *deoxyribonucleic acid*, or in abbreviated form, DNA. This giant molecule contains a coded blueprint in its molecular structure. Indeed, DNA is the universal alphabet of the book of life.

Reductionists would have us believe that DNA is the only immortal part of all organisms, including humans. At the chemical level, life may be just DNA's way of making more DNA. But even though some are inclined to view us as a mere repository for the DNA molecule, we do enjoy a highly distinctive brain that enables us to reflect on our impression of the external world. Today, there is a sense of excitement as we use our intellectual abilities to manipulate and engineer the DNA molecule. Modern DNA technology has so sharpened the precision of molecular analysis that it has found application in virtually all fields of inquiry, from clincal investigations to studies in the evolutionary history of the human species. Indeed, the current body of knowledge of evolution is inextricably entwined with molecular biology.

This updated edition of *Understanding Evolution* is a blend of the time-honored, established body of information and the newer knowledge. To accommodate topics of current interest, all chapters have been extensively revised, and three new chapters and over forty illustrations have been added. You will find information on human gene mapping and DNA sequencing, mitochondrial "Eve," homeotic genes, exon shuffling, and geographical distribution of genes in human populations. We offer no apology for drawing many of the examples from our own species. There is no doubt that we live in an era of profound change. Molecular biologists throughout the world are engaged in a project to provide a blueprint of our total genetic endowment, or technically, our *genome*. The human genome may be thought of as a library of 6 billion letters, which are arranged in code-like fashion to build the human body. As we gain access to the elaborate library and learn how to decipher the code in its entirety, we can then read the approximately 100,000 human genes like an owner's manual. We all share the hope that the manual will be "user-friendly."

Evolution is a central concept in science. The word "evolution" is derived from the Latin *evolutio*, meaning "an unraveling" or "an unfolding." The term suggests gradual change. Broadly speaking, evolution conveys the idea of a continually changing universe as opposed to a universe at a standstill. Organisms throughout life's history have not remained constant but have gradually and endlessly changed. Change is the leitmotif of living things.

The occurrence of biological evolution does not in itself reveal *how* evolution is brought about. An event or phenomenon may be known to us and accepted as true, even though we may not fully understand the forces that determine its existence. Scientists no longer debate that evolution, as a process, has occured. It is in the *explanation* of biological evolution that differences of opinion have arisen. One may challenge an interpretation, but to contest the interpretation is not to deny the existence of the event itself. A widespread fallacy is to discredit the reality of evolution by seizing on points of disagreement concerning the mechanism of evolution.

The rewriting of this edition has been sparked by students who have enjoyed reading the earlier concise renditions. We have sought to write simply yet effectively, without overburdening the reader with excessive details. If simplicity and clarity of presentation of the subject matter have been achieved in this book, it will be due in great measure to listening intently to ever-probing college freshmen in our introductory courses over the many years. The list of *Suggested Readings* at the end of the textbook is intended for the student's enrichment and pleasure. The list includes recent enlightening articles from *Scientific American*, *Natural History*, *Discovery*, and *The New York Times*.

We thank the scores of reviewers for their invaluable comments as we grappled with content and direction. Reviewers of the fifth edition: Jane Aloi, Saddleback College; Albert R. Buckelew, Jr., Bethany College; John Cruzan, Geneva College; Holly Downing, University of Wisconsin, Whitewater; William F. Ettinger, Gonzaga University; Susan Foster, Clark University; David G. Futch, San Diego State University; J. Robert Heckman, Elizabethtown College; Michael J. Patrick, Penn State University, Altoona College; Cathy Schaeff, American University; Sandra Steingraber, Illinois Wesleyan University; Garrison Wilkes, University of Massachusetts, Boston.

We owe a debt of gratitude to our spouses, Lesley and Robin, who quizzically read versions of various chapters for ambiguities. The editorial labors of Marge Kemp, Donna Nemmers, Mary Lee Harms, and other staff members at McGraw-Hill contributed to sharpening the appeal of the presentation. Finally, we acknowledge the enthusiastic support of our colleagues at our respective universities.

E. Peter Volpe
Macon, Georgia
January 1, 1999

Peter A. Rosenbaum
Oswego, New York
January 1, 1999

VARIATION IN POPULATIONS

In the fall of 1958, the folks of a quiet, rural community in the southern part of the United States were startled and dismayed by the large numbers of multilegged frogs in an 85-acre artificial lake on a cotton farm. Widespread newspaper publicity of this strange event attracted the attention of university scientists, curiosity seekers, and gourmets. The lake supported a large population of bullfrogs, estimated at several thousand. Although reports tended to be exaggerated, there were undoubtedly in excess of 350 multilegged deviants. As illustrated in figure 1.1, the extra legs were oddly positioned, but they were unmistakably copies of the two normal hind limbs. Incredibly the extra limbs were functional, but their movements were perceptibly not in harmony with the pair of normal legs. The bizarre multilegged frogs were clumsy and graceless.

All the multilegged frogs appeared to be of the same age, approximately two years old, and of the same generation. These atypic frogs were found only during the one season and were not detected again in subsequent years. The multilegged frogs disappeared almost as dramatically as they had appeared.

History repeated itself in August 1995, almost four decades later, with the appearance, in the wetlands of Minnesota and neighboring states, of many frogs with misshapen hind limbs. In this instance, these multilegged frogs gained instant notoriety, since information and pictures were heralded on the Internet (fig. 1.2). The Minnesota assemblage of variant frogs consisted of several different species. The most severely affected was the highly aquatic mink frog (taxonomically speaking, *Rana septentrionalis*), known for its distinct musky odor. Well over 200 malformed mink frogs were captured at one site in one year. An array of different malformations have been encountered—multiple legs, missing limbs (whole or in part), distorted limbs, missing eyes, and deformed jaws. Disconcertingly, the disfigured frogs in the Minnesota environs have reappeared in successive years—as many as four years in several populations. This differs noticeably from the aforementioned multilegged bullfrogs (*Rana catesbeiana*) that presented themselves in the Mississippi farm locality for only a relatively brief period.

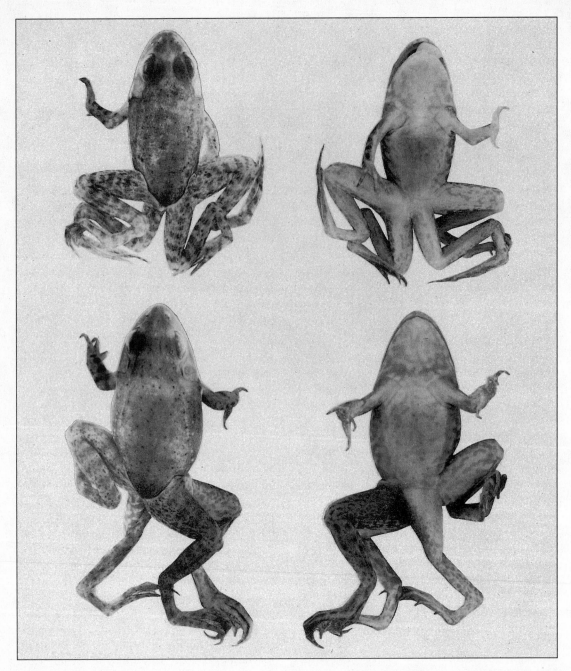

Figure 1.1 Two multilegged frogs, each viewed from the back (dorsal) and front (ventral) surface. These bizarre bullfrogs were discovered in October 1958 in a lake near Tunica, Mississippi. Several hundred frogs with extra hind limbs were found at this locality. How does such an abnormality arise? Two reasonable interpretations are set forth in chapter 1.

(Photographs by E. Peter Volpe, then at Tulane University.)

Figure 1.2 **Dorsal** *(a)* **and ventral** *(b)* **views of a multilegged mink frog** *(Rana septentrionalis)* collected in the wetlands of Minnesota by David Hoppe of the University of Minnesota. Several extra hind limbs protrude from the sides.
Source: Photo by David Hoppe, The University of Minnesota (Morris campus).

Strange and exceptional events of this kind are of absorbing interest, and challenge us for an explanation. How do such oddities arise and what are the factors responsible for their ultimate disappearance in a natural population? In ancient times, bodily deformities evoked reverential awe and inspired some fanciful tales. Early humans constructed a number of myths to explain odd events totally beyond their control or comprehension. One legend has it that when masses of skeletons are revived or reanimated, the bones of different animals often become confused. Another old idea is that grotesquely shaped frogs are throwbacks to some remote prehistoric ancestor. These accounts are, of course, novelistic and illusory. They do, however, reveal the uniquely imaginative capacity of the human mind. Nevertheless, we should seek a completely different cause-and-effect sequence, relying more on our faculty for logical analysis.

ENVIRONMENTAL MODIFICATION

Inspection alone cannot reveal the underlying cause of the multilegged anomaly. We may thoroughly dissect the limbs and describe in detail the anatomy of each component, but no amount of dissection can tell us how the malformation arose. The deformity either was foreordained by heredity or originated from injury to the embryo at a vulnerable stage in development. We shall direct our attention first to the latter possibility and its implications.

Some external factor in the environment may have adversely affected the pattern of development of the hind limb region. The cotton fields around the lake in Mississippi were periodically sprayed with pesticides to combat noxious insects. Purulent airborne pollutants may have settled onto the landscape in Minnesota. The possibility cannot be ignored that excess ultraviolet radiation leaking through a thinning protective ozone layer of the atmosphere caused the limb deformities. In essence, it is not inconceivable that the pesticides, pollutants, and even ultraviolet rays were potent *teratogens*—that is, chemical and physical agents capable of causing marked distortions of normal body parts.

That chemical substances can have detrimental effects on the developing organism is well documented. For example, the geneticist Walter Landauer demonstrated in the 1950s that a wide variety of chemicals, such as boric acid, pilocarpine, and insulin (normally a beneficial hormone), can produce abnormalities of the legs and beaks when injected into chick embryos. This type of finding cannot be dismissed complacently as an instance of a laboratory demonstration without parallel in real life. Indeed, in 1961, medical researchers discovered with amazement amounting to incredulity that a purportedly harmless sleeping pill made of the drug thalidomide, when taken by a pregnant woman, particularly during the second month of pregnancy, could lead to a grotesque deformity in the newborn baby, a rare condition in humans called *phocomelia*—literally, "seal limbs." The arms are absent or reduced to tiny, flipperlike stumps (fig. 1.3).

More than 6,000 thalidomide babies were born in West Germany and at least 1,000 in other countries. The United States was largely spared the thalidomide disaster by the astute scientific sense of Frances O. Kelsey of the U.S. Food Drug Administration: she blocked general distribution of the drug. Despite her efforts, some American women did obtain the drug from European sources. The thalidomide tragedy changed the attitude of the medical profession toward drug-induced malformations. It also awakened the mass media; congenital defects rapidly became a subject of compelling interest to the public.

Thalidomide was banned worldwide in 1962. Strange as it may seem, the world's most infamous drug has resurfaced in recent years. Commercial laboratories have been testing the drug for possible beneficial effects. In September of 1997, the Food and Drug Administration approved the use of thalidomide for the treatment of a severe complication of leprosy, for which no good alternative therapies are presently available. Ongoing research suggests that thalidomide may be useful in the treatment of certain types of cancerous growths (brain tumors) and the wasting syndrome associated with AIDS. There exists gnawing apprehension about the distribution and marketing of the drug.

In different organisms—fish, frogs, mice, and humans—anomalies can be caused by such diverse agents as extremes of temperature, x rays, viruses, drugs, diet deficiencies, and lack of oxygen. There is no longer the slightest doubt that environmental factors may be causal agents of specific defects. A variation that arises as a direct response to some external change in the environment, and not by any change in the genetic makeup of the individual, is called an *environmental modification*.

If the multilegged anomaly in the bullfrog was environmentally induced, we may surmise (as depicted in figure 1.4) that the harmful chemical or other causative factor acted during the sensitive early embryonic stage. Moreover, if some external agent had brought about the abnormality in the bullfrog, this agent must have been effective only once, because the multilegged condition did not occur repeatedly over the years in the Mississippi region. It may be that the harmful environmental factor did not recur, in which case we would not expect a reappearance of the malformation.

Our suppositions could be put to a test by controlled breeding experiments, the importance of which cannot be overstated. Only through breeding tests can the basis of the variation be firmly established. If the anomaly constituted an environmental modification, then a cross of two multilegged bullfrogs would yield all normal

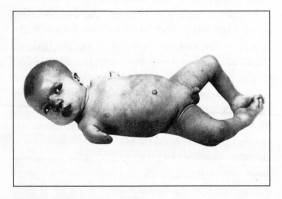

Figure 1.3 Armless deformity in the newborn infant, resulting from the action of thalidomide, a sedative taken by his mother in her second month of pregnancy.

Source: W. Lenz and K. Knapp, "Thalidomide Embryopathy," *Archives of Environmental Health* 5 (1962):100–105. Courtesy of the publisher.

progeny, as illustrated in figure 1.4. In the absence of any disturbing environmental factors, the offspring would develop normal limbs. This breeding experiment was not actually performed; none of the malformed frogs survived to sexual maturity. Nonetheless, we have brought into focus an important biological principle: *environmentally induced (non-genetic) traits cannot be passed on to another generation.*

The elements that are transmitted to the next generation are two minute cells, the egg and the sperm. These two specialized cells are often referred to as *germ* cells, because they are the beginnings, or germs, of new individuals. The germ cells represent the only connecting thread between successive generations. Accordingly, the mechanism of hereditary transmission must operate across this slender connecting bridge. The hereditary qualities of the offspring are established at the time the sperm unites with the egg. The basic hereditary determiners, the *genes,* occur in pairs in the fertilized egg. Each inherited characteristic is governed by at least one pair of genes, with one member of each pair coming from the male parent and the other from the female parent. Stated another way, the sperm and egg each has half the number of genes of the parents. Fertilization restores the original number and the paired condition (fig. 1.4).

The underlying assumption in the cross depicted in figure 1.4 is that the genes that influence the development of the hind limbs are normal ones. Of course, the multilegged parents themselves possess normal genes. It might seem strange that an abnormal character can result from a perfectly sound set of genes. However, normal genes cannot be expected to act normally under all environmental circumstances. A gene may be likened to a photographic negative. A perfect negative (normal gene) may produce an excellent or poor positive print (normal or abnormal trait), depending on such factors (environmental factors) as the quality or concentration of the chemical solutions used in preparing the print. The environment thus affects the *expression* of the

negative (gene), but the negative (gene) itself remains unaffected throughout the making of the print (trait).

We shall learn more about the nature of genes and the mechanism of inheritance in the next chapter. For the moment, the important consideration is that *a given gene prescribes a potentiality for a trait and not the trait itself.* Genes do not act in a vacuum. Genes always act within the conditioning framework of the environment. The materialization of a trait represents the interplay of genetic determinants and environmental factors.

LAMARCKISM

Few people would expect bodily deformities caused by harmful environmental factors to be heritable. And yet, many persons believe that favorable or beneficial bodily changes acquired or developed during one's lifetime are transmitted to the offspring. As a familiar example, athletes who exercised and developed large muscles would pass down their powerful muscular development to their children. This is the famous theory of the *inheritance of acquired characteristics,* or *Lamarckism* (after Jean Baptiste de Lamarck, a French naturalist of the late 1700s and early 1800s). The concept of Lamarckism has no foundation of factual evidence. We know, for example, that a woman who has altered her body by injections of silicone does not automatically pass the alterations on to her daughter. Circumcision is still a requisite in the newborn male, despite a rite that has been practiced for well over 4,000 years. It is sufficient to state that the results of countless laboratory experiments testing the possibility of the inheritance of acquired, or environmentally induced, bodily traits have been negative.

The notion of the inheritance of acquired characteristics became the cornerstone of Lamarck's comprehensive, although incorrect, explanation of evolution. Lamarck's theory is exemplified by a fabled quotation from his book *Philosophie Zoologique* (1809):

The giraffe lives in places where the ground is almost invariably parched and without grass. Obliged to browse on trees it is continually forced to stretch upwards. This habit sustained over long periods of time by every individual of the race has resulted in the forelimbs becoming longer than the hind ones, and the neck so elongated that a giraffe can lift his head to a height of six metres [about 20 feet] without taking his forelimbs off the ground.

Lamarck advocated that the organs of an animal became modified in appropriate fashion in direct response to a changing environment. The various organs became greatly improved through use or reduced to vestiges through disuse. Such bodily modifications, in some manner, could be transferred and impressed on the germ cells to affect future generations. Thus, the whale lost its hind limbs as the consequence of the inherited

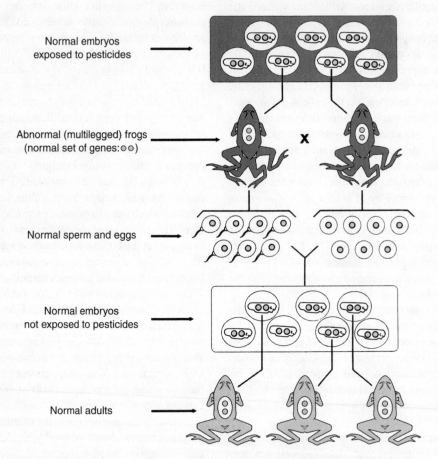

Figure 1.4 Test for the noninheritance of an environmentally induced trait. If the malformation of the hind limbs arose from without by the action of an adverse environmental agent (pesticide) and not from within by genetic change, then the abnormality would *not* be transmitted from parent to offspring. The multilegged parents, although visibly abnormal, are both genetically normal. Normal hind limbs are expected if, during development, the offspring are not exposed to the same injurious environmental agents to which the parents were exposed.

effects of disuse, and the wading bird developed its long legs through generations of sustained stretching to keep the body above the water level. Inheritance was viewed simply as the direct transmission of those superficial bodily changes that arose within the lifetime of the individual owing to use or disuse. Lamarck's views are unacceptable because we know of no mechanism that allows changes in the body to register themselves on the germ cells.

The foregoing considerations bear importantly on our bullfrog population. We have not been able to perform the critical breeding test to prove or disprove that the multilegged trait is a noninheritable modification. However, let us for the moment accept the thesis that the multilegged condition was environmentally induced and examine the implications of this view on the population as a whole. Outwardly, a striking change in the bullfrog population seems to have occurred. But in actuality the composition of the population has remained essentially unaltered, since the basic hereditary materials, the genes, were not affected. In other words, there was no change of evolutionary significance in the population. Evolution can occur *only* where there is heritable variation.

HERITABLE VARIATION

We shall now examine the possibility that the multilegged frogs were genetically abnormal. One or more of their genes may have been impaired. Unlike teratogens, which damage an already conceived offspring and affect only a single generation, deleterious genes are in the germ plasm before conception and may lie dormant for several generations.

We may assume that the multilegged trait is due to an alteration in a single gene, which is transmitted in simple Mendelian fashion. We infer the existence of two alternative forms of the gene—namely, a normal version and a variant version. Geneticists use the Greek word *allele* to denote the normal and variant forms of a gene. In other words, the normal allele of a gene conveys

appropriate instructions, whereas the variant allele of the gene conveys improper information.

We shall postulate that the variant allele for the multilegged condition is *recessive* to the normal allele. That is to say, the expression of the variant allele is completely masked or suppressed by the normal (or *dominant*) allele when the two are present together in the same individual (fig. 1.5). Such an individual, normal in appearance but harboring the detrimental recessive allele, is said to be a *carrier*. The recessive allele may be transmitted without any outward manifestation for several generations, continually being sheltered by its dominant partner. However, as seen in figure 1.5, the deleterious recessive allele ultimately becomes exposed when two carrier parents happen to mate. Those progeny that are endowed with two recessive alleles, one from each parent, are malformed. On the average, one-fourth of the offspring will be multilegged.

Genetic defects transmitted by recessive genes are not at all unusual. Pertinent to the present discussion is an inherited syndrome of abnormalities in humans, known as the Ellis-van Creveld syndrome. Afflicted individuals are disproportionately dwarfed (short limbed), have malformed hearts, and possess six fingers on each hand (fig.1.6). The recessive gene that is responsible for this complex of defects is exceedingly rare. Yet, as shown by geneticist Victor A. McKusick of Johns Hopkins University, the Ellis-van Creveld anomaly occurs with an exceptionally high incidence among the Amish people in Lancaster County, Pennsylvania. The variant recessive allele apparently was present in one member of the original Old Order Amish immigrants from Europe two centuries ago. For a few generations, the harmful allele was passed down unobserved, masked by its normal partner allele. Since 1860, the Ellis-van Creveld deformity has appeared in at least 50 offspring. Ordinarily, it is uncommon for both members of a married couple to harbor the defective recessive allele. However, in the sober religious Amish community, marriages have been largely confined within members of the sect with a resulting high degree of consanguinity. Marriages of close relatives have tended to promote the meeting of two normal, but carrier, parents.

We know very little about the past history or breeding structure of the Mississippi bullfrog population in which the multilegged trait appeared with an exceptionally high frequency. We may suspect that all 350 or more multilegged frogs were derived from a single mating of two carrier parents. In contrast to humans, a single mated pair of frogs can produce well over 10,000 offspring. The similarity in age of the multilegged bullfrogs found in nature adds weight to the supposition that these frogs are members of one generation, and probably of one mating.

Much of the preceding discussion on the multilegged bullfrog is admittedly speculative. However, one aspect is certain: *previously concealed harmful genes are brought to light through the mechanism of heredity.* A trait absent for many generations can suddenly appear without warning. Once a variant character expresses itself, its fate will be determined by the ability of the individual displaying the trait to survive and reproduce in its given environment.

It is difficult to imagine that the grotesquely shaped frogs could compete successfully with their normal kin. However, we shall never know whether or not the multilegged frogs were capable

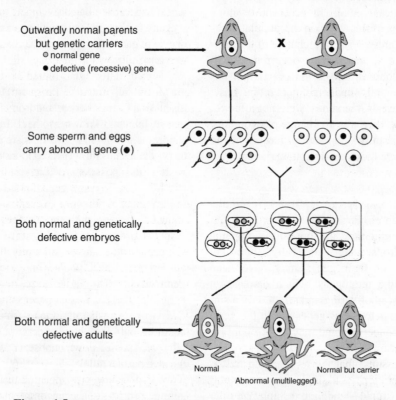

Outwardly normal parents but genetic carriers
  ○ normal gene
  ● defective (recessive) gene

Some sperm and eggs carry abnormal gene (●)

Both normal and genetically defective embryos

Both normal and genetically defective adults

Normal

Abnormal (multilegged)

Normal but carrier

Figure 1.5 Mating of two "carrier" parents resulting in the emergence of multilegged offspring. The multilegged trait is assumed to be controlled by a defective recessive gene. Both parents are normal in appearance, but each carries the defective gene, masked by the normal gene. The recessive gene manifests its detrimental effect when the offspring inherits one abnormal gene from each of its parents.

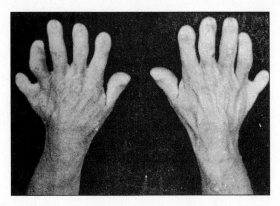

Figure 1.6 Six-digited hands, one of the manifestations of the Ellis-van Creveld syndrome in humans, a rare recessive deformity. Affected offspring generally come from two normal, but carrier, parents, each of whom harbors the abnormal recessive gene.
Courtesy of Victor A. McKusick, Johns Hopkins University.

of contending with the severities of climatic or seasonal changes, or of successfully escaping their predators, or even of actively defending themselves. What we do know is that the multilegged bullfrogs did not survive to reproductive age. Despite diligent searches by many interested investigators in the Mississippi locale, no sexually mature abnormal frogs have been uncovered in the natural population. It seems that this unfavorable variant has been eliminated. In the next chapter, we will explore this situation in light of Darwin's tenet of natural selection.

HYPOTHETICO-DEDUCTIVE REASONING

We have acted as objective and open-minded spectators in observing and interpreting the events surrounding the origin and demise of the multilegged frogs. We formulated a few *hypotheses* or, in the vernacular, educated guesses or hunches. One of the provisional conjectures for the Mississippi assemblage of bullfrogs was that a chance mating of two carrier parents resulted in genetically defective offspring. This hypothesis

permitted the *deduction* that the unfavorable trait would not likely resurface in a subsequent breeding season. In essence, we rendered a hypothesis consistent with the observations, and deduced predictions or consequences from the hypothesis.

We will accept, reject, or modify the hypothesis in accordance with the degree of fulfillment of the predictions. If the hypothesis is to become a scientific explanation, we are obliged to test the deductions by additional observations or by performing experiments. This mode of inquiry is the *hypothetico-deductive* style of reasoning that is the cornerstone of investigative sciences, particularly the natural sciences. This method of establishing explanations for observed phenomena has the goal of developing the habit of scientific thinking, characterized by objectivity, open-mindedness, skepticism, and the willingness to suspend judgment if there is insufficient evidence.

Our prediction was fulfilled for the 1958 population of malformed bullfrogs in the Mississippi area: no such frogs were found again in subsequent breeding seasons. The supposition that a heritable transmission of a faulty gene has been in play in the more recent Minnesota populations of mink frogs is less inviting, particularly in light of a fruitful experiment undertaken by David Hoppe of the University of Minnesota. He examined the outcome of dividing into two batches a cluster of freshly deposited frog eggs from a natural pond where malformed frogs had occurred. One batch was left at its site of collection and subject to the vicissitudes of nature. The second batch was reared in the laboratory in treated water known to be suitable for the growth of tadpoles. Accordingly, the two batches of eggs had the same genetic heritage but were submitted to different environmental experiences. The informative result was that some of the eggs left in nature developed into froglets having extra legs and other malformations, whereas all the laboratory-reared eggs developed into normal froglets. This suggests strongly that some enduring noxious environmental agent, in or around the pond, has persisted in inducing abnormal development of the frogs through the years.

We are not yet justified in accepting any hypothesis as the most probable explanation. If we were able to formulate critical tests that continually substantiate our deductions, our confidence would grow that a particular thesis is the most probable explanation. This explanatory statement would then be regarded as "true" and would become part of the arsenal of scientific knowledge.

Acceptance of a scientific statement is the attainment of a high degree of confidence in its accuracy. But even the most definitive statement is subject to revision. A statement acceptable at one moment of time is still open to possible exception and modification. The natural sciences disallow absolute certainty. Stated another way, there is *no eternal or absolute truth in the natural sciences.* This view offers shallow comfort to those persons who are uncomfortable with, or unaccustomed to, uncertainty.

When a hypothesis is recognized as having far-reaching implications and is continually veri- fied by observation and experimentation, the hypothesis may then attain the status of a *theory.* A theory is, accordingly, a hypothesis sufficiently tested for a scientist to have complete confidence in its probability. A theory is not merely an untest- ed opinion, as implied by the oft-heard recanta- tion: "After all, evolution is *only* a theory!" The theory that life has evolved and is ever changing is founded on as much evidence as the broadly accepted generalization that the earth is round.

Occasionally, but hesitatingly, a theory advances to a law. In the natural sciences, laws have questionable utility. A law tends to connote that a phenomenon is absolutely certain. The famous Mendelian laws of inheritance remain fundamental- ly sound, but they are not, in fact, inviolate. Several dramatic exceptions have recently been uncovered, which Gregor Mendel could not have dared envi- sion. Modern molecular biologists have attached several significant codicils to Mendel's bequest.

Darwinian Scheme of Evolution

More than a century ago, in 1859, Charles Robert Darwin (fig. 2.1) gave the biological world the master key that unlocked all previous perplexities about evolution. His thesis of natural selection can be compared only with such revolutionary ideas as Newton's law of gravitation and Einstein's theory of relativity. The concept of natural selection was set forth clearly and convincingly by Darwin in his monumental treatise *The Origin of Species.* This epoch-making book was the fruition of more than twenty years of meticulously sifting through the voluminous data he had accumulated.

In 1831, Charles Darwin, then 22 years old and fresh from Cambridge University, accepted the post of naturalist, without pay, aboard H.M.S. *Beagle,* a ship commissioned by the British Admiralty for a surveying voyage around the world. Although Darwin was an indifferent student at Cambridge, he did show an interest in the natural sciences. He was an ardent collector of beetles, enjoyed bird watching and hunting, and was an amateur geologist.

It took the *Beagle* nearly five years—from 1831 to 1836—to circle the globe (fig. 2.2). When Darwin first embarked on the voyage, he did not dispute the dogma that every species of organism had come into being at the same time and had remained permanently unaltered. He shared the views of his contemporaries that all organisms had been created about 4000 B.C.—more precisely, at 9:00 A.M. on Sunday, October 23, in 4004 B.C., according to the bold pronouncement of Archbishop James Ussher in the seventeenth century. Darwin had, in fact, studied for the clergy at Cambridge University. But he was to make observations on the *Beagle's* voyage that he could not reconcile with accepted beliefs.

Darwin's quarters on the *Beagle* were cramped, and he took only a few books on board.

Figure 2.1 **Charles Darwin** at the age of 31 (1840), four years after his famous voyage around the world as an unpaid naturalist aboard H.M.S. *Beagle*.

(Courtesy Dept. of Library Services American Museum of Natural History.)

One of them was the newly published first volume of Charles Lyell's *Principles of Geology,* a parting gift from his Cambridge mentor, John Henslow, professor of botany. Lyell challenged the prevailing belief that the earth had been created by a divine plan merely 6,000 years ago. On the contrary, the earth's age could be measured in hundreds of millions of years. Lyell argued that the earth's mountains, valleys, rivers, and coastlines were shaped not by Noah's Flood but by the ordinary action of the rains, the winds, earthquakes, volcanoes, and other natural forces. Darwin was impressed by Lyell's emphasis on the great antiquity of the earth's rocks and gradually came to perceive that the characteristics of organisms, as well as the face of the earth, could change over a vast span of time.

The living and extinct organisms that Darwin observed in the flat plains of the Argentine pampas and in the Galápagos Islands sowed the seeds of Darwin's views on evolution. From old river beds in the Argentine pampas, he dug up bony remains of extinct mammals. One fossil finding was the mas-

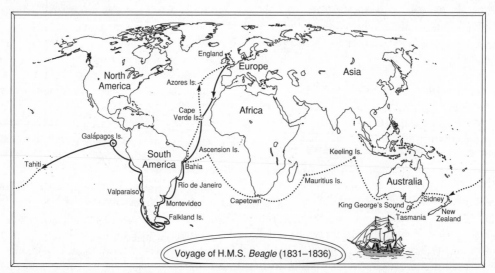

Figure 2.2 **Five-year world voyage of the H.M.S.** *Beagle.* Darwin's observations on this voyage convinced him of the reality of evolution. Particularly impressive to him were the fossil skeletons of mammals unearthed in the Argentine pampas (see fig. 2.3) and the variety of tortoises and birds in the small group of volcanic islands, the Galápagos Islands (see fig. 12.3).

sive *Toxodon,* whose appearance was likened to a hornless rhinoceros or a hippopotamus (fig. 2.3). Another fossil remain that attracted Darwin's attention was the skeleton of *Macrauchenia,* which he erroneously thought was clearly related to the camel because of the structure of the bones of its long neck. Other remarkable creatures were the huge *Pyrotherium,* resembling an elephant, and the light and graceful single-toed *Thoatherium,* rivaling the horse. The presence of *Thoatherium* testified that a horse had been among the ancient inhabitants of the continent. It was the Spanish settlers who reintroduced the modern horse, *Equus,* to the continent of South America in the sixteenth century. Darwin marveled that a native horse should have lived and disappeared in South America. This was one of the first indications that species gradually became modified with time and that not all species survived through the ages.

When Darwin collected the remains of giant armadillos and sloths on an Argentine pampa, he reflected on the fact that, although they clearly belonged to extinct forms, they were constructed on the same basic plan as the small living armadillos and sloths of the same region. This experience started him thinking of the fossil sequence of a given animal species through the ages and the causes of extinction. He wrote: "This wonderful relationship in the same continent between the dead and the living will, I do not doubt, hereafter throw more light on the appearance of organic beings on our earth, and their disappearance from it, than any other class of facts."

What Darwin had realized was that living species have ancestors. This fact is commonplace now, but it was a revelation then. On traveling from the north to the south of South America, Darwin observed that one species was replaced by similar, but slightly different, species. In the southern part of Argentina, Darwin caught a rare species of ostrich that was smaller and differently colored from the more northerly, common American ostrich, *Rhea americanus.* This rare species of bird was later named after him, *Rhea*

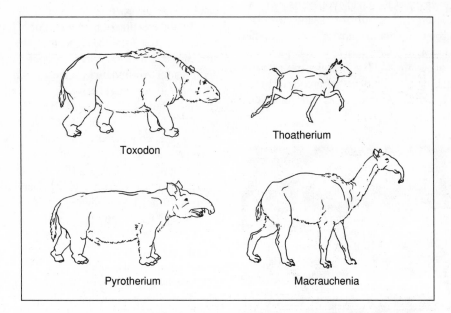

Figure 2.3 Curious hoofed mammals (ungulates) that flourished on the continent of South America some 60 to 70 million years ago and have long since vanished from the scene. Bones of these great mammals were found by Darwin on the flat treeless plains of Argentina.

darwini. It was scarcely imaginable to Darwin that several minor versions of a species would be created separately, one for each locality. It appeared to Darwin that species change not only in time but also with geographical distance. He later wrote: "It was evident that such facts could only be explained on the supposition that species gradually became modified; and the subject haunted me."

In the Galápagos Islands, Darwin's scientific curiosity was sharply prodded by the many distinctive forms of life. The Galápagos consist of an isolated cluster of rugged volcanic islands in the eastern Pacific, on the equator about 600 miles west of Ecuador. One of the most unusual animals is the giant land-dwelling tortoise, which may weigh as much as 550 pounds, grow to 6 feet in length, and attain an age of 200 to 250 years (fig. 2.4). The Spanish name for tortoise, *galápago,* gives the islands their name. Darwin noticed that the tortoises were clearly different from island to island, although the islands were only a few miles apart. Darwin reasoned that, in isolation, each population had evolved its own distinctive features. Yet, all the island tortoises showed basic resemblances not only to each other but also to relatively large tortoises on the adjacent mainland of South America. All this revealed to Darwin that the island tortoises shared a common ancestor with the mainland forms. The same was true of a group of small black birds, known today as *Darwin's finches.* Darwin observed that the finches were different on the various islands, yet they were obviously closely related to each other. Darwin reasoned that the finches were derived from an ancestral stock that had emigrated from the mainland to the volcanic islands and had undergone profound changes under the different conditions of the individual islands. Apparently, a single ancestral group could give rise to several different varieties or species.

MECHANISM OF EVOLUTION

When Darwin left England in 1831, he had accepted the established theory that different species had arisen all at once and remained unchanged and constant. His scientific colleagues did not deny the presence of fossil remains of species no longer extant. Such extinct species were thought to represent organisms that had been periodically annihilated by sudden and violent catastrophes. Each cataclysmic destruction was said to be accompanied by the creation of new forms of life, each distinct from the fossilized species. The theory of successive episodes of catastrophic destruction and creation, formulated by the French anatomist Georges Cuvier, was accepted by most eighteenth-century scholars.

On his return to England in October 1836, Darwin was convinced of the truth of the idea of *descent with modification*—that all organisms, including humans, are modified descendants of previously existing forms of life. Darwin's observations during his voyage assured him that animals and plants are not the products of special creation, as Victorian society believed. We can discern two stages in the development of Darwin's thought: the first was the realization that the world of organisms is not fixed and unchangeable; the second was an explanation of the process of evolutionary change.

At home, Darwin assembled data relevant to the mechanism of evolution. He became intrigued by the extensive domestication of animals and

Figure 2.4 **The giant tortoise,** *Geochelone elephantopus,* found only on the Galápagos Islands.
Source: ©Tess & David Young/Tom Stack & Assoc.

plants brought about by conscious human efforts. Throughout the ages, humans have been a powerful agent in modifying wild species of animals and plants to suit our needs and whims. Through careful breeding programs, we determine which characteristics or qualities will be incorporated or discarded in our domesticated stocks. By conscious *selection,* we perfected the toylike Shetland pony, the Great Dane dog, the sleek Arabian race horse, and vast numbers of cultivated crops and ornamental plants.

Many clearly different domestic varieties have evolved from a single species through our efforts. As an example, we may consider the many varieties of domestic fowl (fig. 2.5), all of which are derived from a single wild species of red jungle fowl *(Gallus gallus)* present at one time in northern India. The red jungle fowl dates back to 2000 B.C. and has been so modified as to be nonexistent today. The variety of fowls perpetuated by fowl fanciers ranges from the flamboyant ceremonial cocks (the Japanese *Onaga-dori*) to the leghorns, bred especially to deposit spotless eggs. In the plant world, horticulturists have evolved the astonishing range of begonias shown in figure 2.6. The wild begonia, from which most of the modern varieties have been bred, was found in 1865 and still exists today in the Andes.

In the late 1830s, Darwin attended the meetings of animal breeders and intently read their publications. Animal breeders were conversant with the variability in their pet animals and utilized the

Figure 2.5 Evolution under domestication. A variety of domestic fowls have evolved from a single wild species (now extinct) as a result of the continual practice by humans of artificial selection.

Black-tailed Japanese bantam

Full-sized and bantam white leghorn

Crested white duck

Sebastopol goose

Onaga-dori cock

Frizzled sultan rooster

Birchen game bantam

Araucana

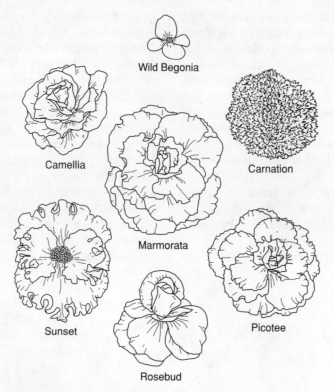

Wild Begonia

Camellia

Carnation

Marmorata

Sunset

Picotee

Rosebud

Figure 2.6 Artificial selection. In the hands of skilled horticulturists, several varieties of the begonia have been established.

technique of *artificial selection.* That is, the breeders selected and perpetuated those variant types
that interested them or seemed useful to them. The
breeders, however, had only vague notions as to
the origin, or inheritance, of the variable traits.

Darwin acknowledged the unlimited variability in organisms but was never able to explain satisfactorily how a variant trait was inherited. He
and other naturalists were unaware of Gregor
Mendel's contemporaneous discovery. Rather,
Darwin followed Lamarck in assuming that biological variation is chiefly conditioned by direct
influences of the environment on the organism. He
believed that the changes induced by the environment became inherited; that bodily changes in the
parents leave an impression or mark on their germ
cells, or gametes. Darwin thought that all organs
of the body sent contributions to the germ cells in

the form of particles, or *gemmules.* These gemmules, or minute delegates of bodily parts, were
supposedly discharged into the blood stream and
became ultimately concentrated in the gametes. In
the next generation, each gemmule reproduced its
particular bodily component. The theory was
called *pangenesis,* or "origin from all," since all
parental bodily cells were supposed to take part in
the formation of the new individual.

Having explained the origin of variation
(although incorrectly), Darwin wondered how
artificial selection (a term familiar then only to
animal breeders) could be carried on in nature.
There was no breeder in nature to pick and
choose. In 1837, Darwin wrote: "How selection
could be applied to organisms living in a state of
nature remained a mystery to me." Slightly more
than a year later, Darwin found the solution. In his

autobiography, Darwin explained: "In October 1838, that is fifteen months after I had begun my systematic enquiry, I happened to read for amusement *Malthus on Population* . . . at once it struck me that under these circumstances favourable variations would tend to be preserved and unfavourable ones destroyed. The result of this would be the formation of a new species." The circumstances mentioned by Darwin were those associated with the assertions of an English clergyman, Thomas Robert Malthus, that population is necessarily limited by the means of subsistence.

DARWIN'S NATURAL SELECTION

Malthus's writings provided the germ for Darwin's theory of *natural selection.* In his famous *An Essay on the Principle of Population,* Malthus expressed the view that the reproductive capacity of humankind far exceeds the food supply available to nourish an expanding human population. Humans compete among themselves for the necessities of life. This unrelenting competition engenders vice, misery, war, and famine. It thus occurred to Darwin that competition exists among all living things. Darwin then envisioned that the "struggle for existence" might be the means by which the well-adapted individuals survive and the ill-adjusted are eliminated. Darwin was the first to realize that perpetual selection existed in nature in the form of *natural selection.* In natural selection, as contrasted to artificial selection, the animal breeder or horticulturist is replaced by the conditions of the environment that prevent the survival and reproduction of certain individuals. The process of natural selection occurs without a conscious plan or purpose. Natural selection was an entirely new concept, and Darwin was its principal proponent.

It was not until 1844 that Darwin developed his idea of natural selection in an essay, but not for publication. He showed the manuscript to the geologist Charles Lyell, who encouraged him to prepare a book. Darwin still took no steps toward publishing his views. It appears that Darwin might

not have prepared his famous volume had not a fellow naturalist, Alfred Russel Wallace (in the Dutch East Indies), independently conceived of the idea of natural selection. Wallace had spent many years exploring and collecting in South America and the West Indies. Wallace was also inspired by reading Malthus's essay, and the idea of natural selection came to him in a flash of insight during a sudden fit of malarial fever. In June of 1858, Wallace sent Darwin a brief essay on his views. The essay was entitled *On the Tendencies of Varieties to Depart Indefinitely from the Original Type.* With the receipt of this essay, Darwin was then induced to make a statement of his own with that of Wallace.

Wallace's essay and a portion of Darwin's manuscript, containing remarkably similar views, were read simultaneously before the Linnaean Society in London on July 1, 1858. The joint reading of the papers stirred little interest. Darwin then labored for eight months to compress his voluminous notes into a single book, which he modestly called "only an Abstract." Wallace shares with Darwin the honor of establishing the mechanism by which evolution is brought about, but it was the monumental *The Origin of Species,* with its impressive weight of evidence and argument, that left its mark on humankind. The full title of Darwin's treatise was *On the Origin of Species by Means of Natural Selection, or the Preservation of Favoured Races in the Struggle for Life.* The first edition, some 1,250 copies, was sold out on the very day it appeared, November 24, 1859. The book was immediately both acidly attacked and effusively praised. Today, *The Origin of Species* remains the one book to be read by all serious students of nature.

The idea of evolution—that organisms change—did not originate with Darwin. There were many before him—notably Lamarck, Georges Buffon, and Charles Darwin's grandfather, Erasmus Darwin—who recognized or intimated that animals and plants have not remained unchanged through time but are continuously changing. Indeed, early Greek philosophers, who wondered about everything, speculated on the gradual progression of life from simple to complex. It was reserved for Darwin to remove the doctrine of evolution from the domain of speculation. Darwin's outstanding achievement was his discovery of the principle of natural selection. In showing *how* evolution occurs, Darwin endeavored to convince skeptics that evolution *does* occur.

As a whole, the principle of natural selection stems from three important observations and two deductions that logically follow from them. The first observation is that all living things tend to increase their numbers at a prolific rate. A single oyster may produce as many as 100 million eggs at one spawning; one tropical orchid may form well over 1 million seeds; and a single salmon can deposit 28 million eggs in one season. It is equally apparent (the second observation) that no group of organisms swarms uncontrollably over the surface of the earth. In fact, the actual size of a given population of any particular organism remains relatively constant over long periods of time. If we accept these readily confirmable observations, the conclusion necessarily follows that not all individuals that are produced in any generation can survive. There is inescapably in nature an intense "struggle for existence."

Darwin's third observation was that individuals in a population are not alike but differ from one another in various features. That all living things vary is indisputable. Those individuals endowed with the most favorable variations, concluded Darwin, would have the best chance of surviving and passing their favorable characteristics on to their progeny. This differential survival, or "survival of the fittest," was termed *natural selection.* It was the British philosopher Herbert Spencer who proposed the expression "survival of the fittest," which Darwin accepted as equivalent to natural selection. A conceptual model of Darwin's tenet of natural selection is presented in figure 2.7.

Darwin's book appeared at a time when England embraced the Victorian economic doctrine of *laissez-faire.* This French phrase expresses the notion of letting people do what they choose. As used in economics, the doctrine deprecates any form of governmental interference intended to regulate labor, manufacturing, commerce, and financial institutions. The Victorian era was one of fierce competition among manufacturers, merchants, and bankers. Herbert Spencer advocated that Darwin had discovered not merely the laws of biological evolution but also those governing human societies. Spencer gave science's blessing to the selection

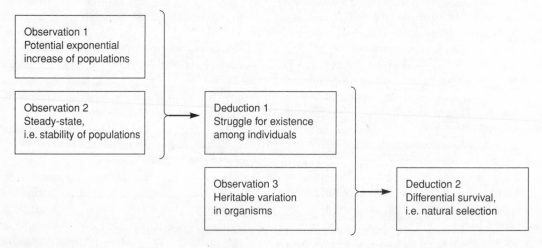

Figure 2.7 **Darwin's concept of natural selection.**

process that operated in the ruthless English society where only the fittest could survive the intolerable working and living conditions. Victorian England was also staunchly biblical. Science was linked in many devout minds with atheistic thinking, and Darwin's book was denounced from fundamentalist pulpits. The Victorian press assailed Darwin for his arrogance in removing humans from an exalted position in the world of life (fig. 2.8).

DIFFERENTIAL REPRODUCTION

The survival of favorable variants is one facet of the Darwinian concept of natural selection. Equally important is the corollary that unfavorable variants do not survive and multiply. Consequently, natural selection necessarily embraces two aspects, as inseparable as the two faces of the same coin: the negative (elimination of the unfit) and the positive

Figure 2.8 Cartoon in Punch's Almanac (1882). The cartoon portrays the transition of worms into apes and gnomes, culminating in a gentleman with sartorial splendor paying facetious homage to Charles Darwin.

Source: Punch's Almanac

(perpetuation of the fit). In its negative role, natural selection serves as a conservative or stabilizing force, pruning out the aberrant forms from a population.

The superior, or fit, individuals are popularly extolled as those who emerge victoriously in brutal combat. Fitness has often been naively confused with physical, or even athletic, prowess. This glorification is traceable to such seductive catch phrases as the "struggle for existence" and the "survival of the fittest." But what does fitness actually signify?

The true gauge of fitness is not merely survival but also the organism's capacity to leave offspring. An individual must survive to reproduce, but not all individuals that survive do, or are able to, leave descendants. We have seen that the multilegged bullfrogs were not successful in propagating themselves. They failed to make a contribution to the next or succeeding generations. Therefore, they were unfit. Hence, individuals are biologically unfit if they leave no progeny. They are also unfit if they do produce progeny, none of whom survives to maturity. The less spectacular normal-legged frogs did reproduce, and to the extent that they are represented by descendants in succeeding generations, they are the fittest. *Fitness, therefore, is measured as reproductive effectiveness.* Natural selection can thus be thought of as *differential reproduction,* rather than differential survival.

EVOLUTION DEFINED

Any given generation is descended from only a small fraction of the previous generation. It should be evident that the genes transmitted by those individuals who most successfully reproduce will predominate in the next generation. Because of unequal reproductive capacities of individuals with different hereditary constitutions, the genetic characteristics of a population become altered each successive generation. This is a dynamic process that has occurred in the past, occurs today, and will continue to occur as long as heritable variation and differing reproductive abilities exist. Under these circumstances, the composition of a population can never remain constant. This, then, is evolution— *changes in the genetic composition of a population with the passage of each generation.*

The outcome of the evolutionary process is adaptation of an organism to its environment. Many of the structural features of organisms are marvels of construction. It is, however, not at all remarkable that organisms possess particular characteristics that appear to be precisely and peculiarly suited to their needs. This is understandable because the individuals who leave the most descendants are most often those who are best equipped to cope with special environmental conditions. In other words, the more reproductively fit individuals tend to be those who are better adapted to the environment.

An important consideration is that evolution is a property of populations. A population is not a mere assemblage of individuals; it is a breeding community of individuals. Moreover, it is the *population* that evolves in time, not the *individual.* An individual cannot evolve, for he or she survives for only one generation. An individual, however, can ensure continuity of a population by leaving offspring. Life is perpetuated only by the propagation of new individuals who differ in some degree from their parents. Natural selection is a term serving to inform us that some individuals leave more offspring than others. Natural selection is not purposeful or guided by a specific aim; it does not seek to attain a specific end. Natural selection is solely the consequence of the differences between individuals with respect to their capacity to produce progeny.

Throughout the ages, appropriate adaptive structures have arisen as the result of gradual changes in the hereditary endowment of a population. Admittedly, past events are not amenable to direct observation or experimental verification. There are no living eyewitnesses of very distant events. So, the process of evolution in the past must be inferred. Nevertheless, we may be confident that the same evolutionary forces we witness in operation today have guided evolution in the past. The basis for this confidence is imparted in the chapters ahead.

3

HERITABLE VARIATION

Darwin recognized that the process of evolution is inseparably linked to the mechanism of inheritance. But he could not explain satisfactorily how a given trait is transmitted from parent to offspring, nor could he account adequately for the sudden appearance of new traits. Unfortunately, Darwin was unaware of the great discovery in heredity made by a contemporary, the unpretentious Austrian monk and mathematics teacher, Gregor Johann Mendel.

Mendel carried out carefully controlled breeding experiments on the common pea plant in his small garden at the monastery of Saint Thomas at Brünn in Moravia (now Brno, Czechoslovakia). He worked with large numbers of plants so that statistical accuracy could be assured. In 1866, Mendel's results appeared in a publication of the Brünn Society of Natural History. In the small compass of 44 printed pages, one of the most important principles of nature ever discovered is clearly revealed. Yet, his paper elicited little interest or comment. No one praised or disputed Mendel's findings. It is likely that the nineteenth-century biologists were baffled by Mendel's mathematical,

or statistical, approach to the study of hybridization. Botany and mathematics were strange bedfellows for that period.

Mendel's valuable contribution lay ignored or unappreciated until 1900, sixteen years after his death. The rediscovery of Mendel's obscure manuscript at the beginning of the twentieth century ushered in the science of genetics and ultimately led to our current understanding of heritable variation.

MENDEL'S PRINCIPLE OF SEGREGATION

Mendel's profound inference was that traits are passed on from parent to offspring through the gametes, in specific discrete units, or *factors*. The individual factors do not blend and do not contaminate one another. The hereditary factors of the parents can, therefore, be reassorted in varied combinations in different individuals at each generation. Today, Mendel's hereditary units are called *genes*. The almost infinite diversity that exists among individuals is attributable largely to the shuffling of tens of thousands of discrete genes that occurs in sexual reproduction.

Mendel reasoned that an individual has two copies of each gene, having received one copy of the gene from the gamete of the female parent and the other copy from the gamete of the male parent. Today, we refer to the contrasting forms of a given gene as *alleles*. Expressed in another way, all alternative sequences of a gene, including the normal and all variant sequences, are called alleles. In the most elementary case of inheritance, the trait is governed by a single gene that exists in two allelic forms—the normal (or typical) allele and the variant (or mutant) allele. Albinism in humans may be taken as an example of a trait under the simple control of a pair of alleles.

The word *albino* is derived from the Latin *albus*, meaning "white," and refers to the paucity or absence of pigment (melanin) in the skin, hair, and eyes. The skin is often very light ("milk white"), and the hair whitish yellow. The eyes appear pinkish because the red blood vessels give the otherwise colorless iris a rosy cast. Contrary to popular opinion, albinos are neither blind nor mentally retarded. They do have poor vision, are acutely sensitive to sunlight, and are prone to life-threatening skin cancer. In some cultures, as among the natives of the Archipelago Coral Islands, albinos are believed to have a mesmeric aura (fig. 3.1).

The absence of melanin is the consequence of a variant allele. Specifically, albinism arises only when the individual is endowed with two variant alleles, one from each parent. A cardinal fact of inheritance is that each gamete (egg or sperm) contains *one, and only one,* member of a pair of alleles. Thus, when the egg and sperm unite, the two alleles are brought together in the new individual. During the production of the gametes, the members of a pair of alleles separate, or *segregate,* from each other. It bears emphasizing that a given gamete can carry only one allele but not both. This fundamental concept that only one member of any pair of alleles in a parent is transmitted to each offspring is Mendel's first principle—the *principle of segregation.*

As seen in figure 3.2, there are nine possible matings with respect to the pair of alleles determining the presence or absence of melanin. If both parents are perfectly normal—that is, each possesses a pair of normal alleles—then all offspring will be normally pigmented (cross 1). When both parents are albino and accordingly each has a pair of variant genes, all offspring will be afflicted with albinism (cross 2). The persons involved in either of these two crosses are said to be *homozygous,* since they each possess a pair of similar, or like, alleles. The normal individuals in cross 1 are homozygous for the normal allele, and the albino individuals in cross 2 are homozygous for the variant allele.

Looking at cross 3, we notice that one of the parents (the female) carries an unlike pair of alleles: one member of the pair is normal; its partner allele is variant. This female parent, possessing unlike alleles, in *heterozygous.* This parent has normal coloration, which leads us to deduce that the expression of the variant allele is completely masked or suppressed by the normal allele. In genetical parlance, we say that the normal allele is *dominant* over its partner, the *recessive* allele. The dominant allele is customarily symbolized by a capital letter (in our case, A); the recessive allele, by a corresponding small letter *(a).* The heterozygous parent is depicted as *Aa*, and is referred to as *heterozygote* or a *carrier.*

When the heterozygote *(Aa)* mates with a dominant homozygote *(AA),* all the offspring will be outwardly normal, but half the offspring will be carriers *(Aa)* like one of the parents (cross 3). The ratio of *AA* to *Aa* offspring may be expressed as 1/2 *AA* : 1/2 *Aa* or simply 1 *AA* : 1*Aa*. Cross 4 differs from cross 3 only in that the male parent is the carrier *(Aa)* and the female parent is the dominant homozygote *(AA).* The outcome is the same: all children are normal in appearance; but half of them are carriers. Thus, with respect to albinism, the males and females are equally likely to have or to transmit the trait. (There are traits, such as hemophilia and red-green color blindness, in which transmission is influenced by the sex of the parents.)

Figure 3.1 Albino person showing the complete absence of pigment in the eyes, hair, and skin. Natives of the Archipelago Coral Islands (100 miles east of the New Guinea mainland) are said to take care of their "pure white" relatives who can hardly see in the glare of the tropical sun.
(Wide World Photos)

All offspring will be carriers from the mating (crosses 5 and 6) of a dominant homozygote *(AA)* and an albino parent *(aa)*. Examining crosses 5 and 6 (fig. 3.2), we again notice that males and females are equally affected. The mating of a heterozygous person *(Aa)* and an albino *(aa)* gives rise to two types of progeny, in equal numbers (crosses 7 and 8). Half the offspring are normal but carriers *(Aa);* the remaining half are albino *(aa)*.

The outcome of the last of the crosses (no. 9) in figure 3.2 is comprehensible in terms of Mendel's law of segregation. Both parents are normal in appearance, but each carries the variant allele *(a)* masked by the normal allele *(A)*. The male parent *(Aa)* produces two kinds of

sperm cells; half the sperm carry the *A* allele; half carry the *a* allele. The same two kinds occur in equal proportions among the egg cells. Each kind of sperm has an equal chance of meeting each kind of egg. The random meeting of gametes leads to both normal and albino offspring. One-fourth of the progeny are normal and completely devoid of the recessive allele *(AA);* one-half are normal but carriers like the parents *(Aa);* and the remaining one-fourth exhibit the albino anomaly *(aa)*. The probabilities of occurrence are the same as those illustrated by the tossing of a coin. If two coins are repeatedly tossed at the same time, the result to be expected on the bases of pure chance is that

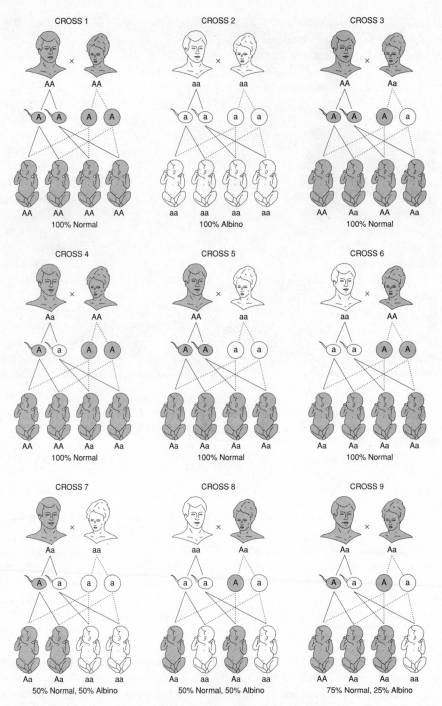

Figure 3.2 **The segregation** of a single pair of alleles. The nine possible marriages with respect to the recessive trait for albinism show the operation of Mendel's principle of segregation.

the combination "head and head" will fall in one-fourth of the cases, the combination "head and tail" in one-half of them, and the combination "tail and tail" in the remaining one-fourth. The ratio of genetic constitutions among the offspring may be expressed as 1/4 *AA* : 1/2 *Aa* : 1/4 *aa,* or 1 *AA* : 2 *Aa* : 1 *aa.*

Geneticists continually have to make a distinction between the external, or observable, appearance of the organism and its internal hereditary constitution. They call the former the *phenotype* (visible type) and the latter the *genotype* (hereditary type). In our last cross (no. 9), there are three different genotypes: *AA, Aa,* and *aa.* The dominant homozygote *(AA)* cannot be distinguished by inspection from the heterozygote *(Aa).* Thus, *AA* and *Aa* genotypes have the same phenotype. Accordingly, on the basis of phenotype alone, the progeny ratio is 3/4 normal : 1/4 albino, or 3 normal : 1 albino.

If we disregard the sexes of the parents, the nine crosses in figure 3.2 become reducible to six possible types of matings. The six possible crosses and their offspring are set forth in table 3.1.

APPLICATION OF MENDELIAN PRINCIPLES

Our interpretation of the mode of inheritance of the multilegged trait in the bullfrog was based on Mendel's principles. In chapter 1, we postulated a cross between two normal, but heterozygous, frogs (see fig. 1.5). The expectation for multilegged offspring from two heterozygous parents is 25 percent, in keeping with Mendel's 3:1 ratio. This cross, using appropriate genetic symbols, is shown graphically by "checkerboard" analysis in figure 3.3. The convenient checkerboard method of analysis was devised by the British geneticist R.C. Punnett. The eggs and sperm are listed separately on two different sides of the checkerboard, and each square represents an offspring that arises from the union of a given egg cell and a given sperm cell. The "reading" of the squares discloses the classical Mendelian phenotypic ratio of 3:1.

It should be understood that the 3:1 phenotypic ratio resulting from the mating of two heterozygous persons, or the 1:1 ratio from the mating of a heterozygote and a recessive person, are

TABLE 3.1	**Simple Recessive Mendelian Inheritance, Involving a Single Pair of Alleles (Normal Pigmentation vs. Albinism)**

Mating Types		Gametes				Offspring	
		First Parent		Second Parent			
Genoptypes	Phenotypes	50%	50%	50%	50%	Genotypes	Phenotypes
AA × *AA*	Normal × Normal	*A*	*A*	*A*	*A*	100% *AA*	100% Normal
AA × *Aa*	Normal × Normal	*A*	*A*	*A*	*a*	50% *AA* 50% *Aa*	100% Normal
Aa × *Aa*	Normal × Normal	*A*	*a*	*A*	*a*	25% *AA* 50% *Aa* 25% *aa*	75% Normal 25% Albino
AA × *aa*	Normal × Albino	*A*	*A*	*a*	*a*	100% *Aa*	100% Normal
Aa × *aa*	Normal × Albino	*A*	*a*	*a*	*a*	50% *Aa* 50% *aa*	50% Normal 50% Albino
aa × *aa*	Albino × Albino	*a*	*a*	*a*	*a*	100% *aa*	100% Albino

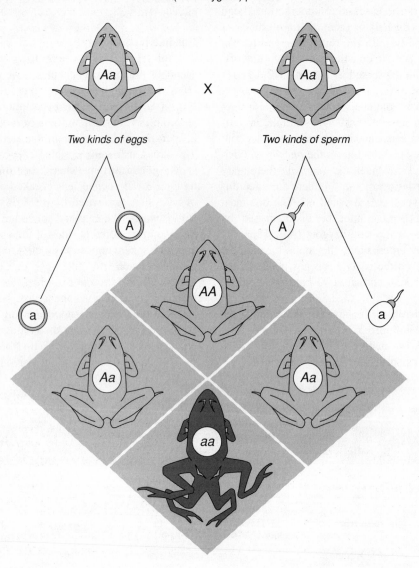

Normal (heterozygous) parents

$^3/_4$ Normal : $^1/_4$ Multilegged

Figure 3.3 Cross of two heterozygous carrier frogs. One-fourth of the offspring are normal and completely devoid of the recessive allele *(AA)*; one-half are normal but carriers, like the parents *(Aa);* and the remaining one-fourth exhibit the multi-legged anomaly *(aa).*

expectations based on probability, and are not invariable outcomes. The production of large numbers of offspring increases the probability of obtaining, for example, the 1:1 progeny ratio, just as many tosses of a coin improve the chances of approximating the 1 head: 1 tail ratio. If a coin is tossed only two times, a head on the first toss is not invariably followed by a tail on the second toss. In like manner, if only two offspring are produced from a marriage of heterozygous *(Aa)* and recessive *(aa)* parents, it should not be thought that one normal offspring is always accompanied by one recessive offspring. With small numbers of progeny, as is characteristic of humans, any ratio might arise in a given family.

Stated another way, the 3:1 and 1:1 ratios reveal the risk or odds of a given child having the particular trait involved. If the first child of two heterozygous parents is an albino, the odds that the second child will be an albino remain 1 out of 4. These odds hold for each subsequent child, irrespective of the number of previously affected children. Each birth is an entirely independent event. If this still appears puzzling, consider once again the tossing of a coin. The first time a coin is tossed, the chance of obtaining either a head or a tail is 1 in 2. Whether the toss is repeated immediately or nine months later, the chance of the coin falling head or tail is still 1 in 2, or 50 percent.

CHROMOSOMAL BASIS OF HEREDITY

Each organism has a definite number of chromosomes, ranging in various species from 2 to more than 1,000. Humans, for example, have 46 microscopically visible, threadlike chromosomes in the nucleus of every body cell (fig. 3.4). Not only are the numbers of chromosomes constant for a given species, but the shapes and sizes of the chromosomes in a given species differ sufficiently for us to recognize particular chromosomes and to distinguish one kind from the other. An important consideration is that chromosomes in a zygote

occur in pairs. The significance of the double, or paired, set of chromosomes cannot be overstated: *each parent contributes one member of each pair of chromosomes to the offspring.*

The chromosomes can be systematically arranged in a sequence known as a *karyotype* (fig. 3.4). In preparing the karyotype, the chromosomes are paired and classified into groups according to their length. There are 22 matching pairs of human chromosomes that are called *autosomes.* The autosome pairs are numbered 1 to 22 in descending order of length, and further classified into seven groups, designated by capital letters, A through G. In addition, there are two sex chromosomes, which are unnumbered. As seen in figure 3.4, the male has one X chromosome and one unequal-sized Y chromosome. The X is of medium size, and the Y is one of the smallest chromosomes of the complement. The human female has two X chromosomes of equal size and no Y chromosome. Thus, the complement of 46 human chromosomes comprises 22 pairs of autosomes plus the sex chromosome pair, XX in normal females and XY in normal males. In current nomenclature, the female is described as "46,XX" and the male, "46,XY." It is of peculiar interest that the long arm of the Y chromosome varies in length in normal individuals. About 3 percent of normal males exhibit variation in the long arms of the Y with no apparent influence on fertility or virility.

MITOSIS AND MEIOSIS

The formation of new cells by *mitosis* is a continual process in the human body, although certain cells divide more often than others, and some not at all. Mitosis is the means by which two new cells receive identical copies of the chromosome complement of the pre-existing cell (fig. 3.5). If gametes were to be produced by the usual mitotic process, and, accordingly, if they had the same number of chromosomes as body cells, then the union of sperm and egg would necessarily double the number of chromosomes in each generation.

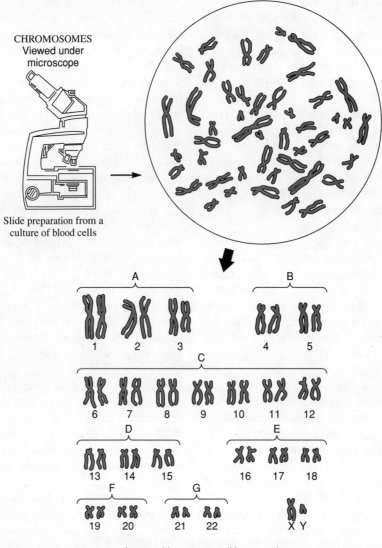

CHROMOSOMES
Viewed under
microscope

Slide preparation from a
culture of blood cells

A   B

1 2 3   4 5

C

6 7 8 9 10 11 12

D   E

13 14 15   16 17 18

F   G

19 20   21 22   X Y

Arranged in sequence (Karyotype)

Figure 3.4 Normal complement of 46 chromosomes prepared from a culture of
blood cells of a human male. Each metaphase chromosome consists of two strands
(chromatids) joined at the centromere. When the chromosomes are arranged in a
sequence (karyotype), the male set consists of 23 pairs, including the XY.

Such progressive increase is forestalled, however, by a special kind of division that guarantees
that the gametes contain *half* the number of chromosomes characteristic of body cells. This dis-

tinctive type of division is referred to as *meiosis.*
The term "meiosis" is derived from the Greek
word *meion,* meaning "to make smaller." Meiosis
is thus a diminution process, referring specifical-

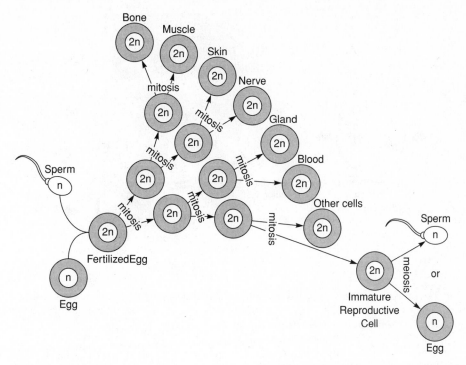

Figure 3.5 **Mitosis** ensures that each new cell of the developing organism has a complete set of the chromosomes (diploid, or 2*n*) found in the fertilized egg cell. Through the process of meiosis, gametes (egg and sperm) come to possess only half the number of chromosomes (haploid, or 1*n*).

ly to a reduction in the number of chromosomes. In humans, the number of chromosomes is reduced from 46 to 23 during the formation of gametes.

Somatic (or body) cells, possessing twice the number of chromosomes as gametes, are said to be *diploid* (Greek, *diploos,* "twofold" or "double"). In contrast, the gametic number of chromosomes is *haploid* (Greek, *haploos,* "single") or, as some investigators prefer, *monoploid* (Greek, *monos,* "alone"). This can be expressed in abbreviated form by saying that body cells contain 2*n* chromosomes and germ cells contain 1*n* chromosomes, where *n* stands for a definite number. As earlier mentioned, a given chromosome has a mate that is identical in size and shape. The two identical members of a pair are called *homolo-*

gous chromosomes, or simply *homologues.* With this orientation, we can explore the mechanism of meiosis.

Meiosis is not a single division but rather two successive ones, the first and second meiotic divisions. Figure 3.6 reveals the essential features in a simplified "stick" representation involving merely one pair of homologous chromosomes. The process begins with each member of the pair duplicating itself to form two sister strands (technically, called *chromatids*). Then, an event that is *unique* to meiosis occurs: the homologous chromosomes actually come to lie alongside one another. The mutual attraction, or pairing, of homologues is known as *synapsis.*

Since two homologous chromosomes pair, and since each chromosome of a homologous

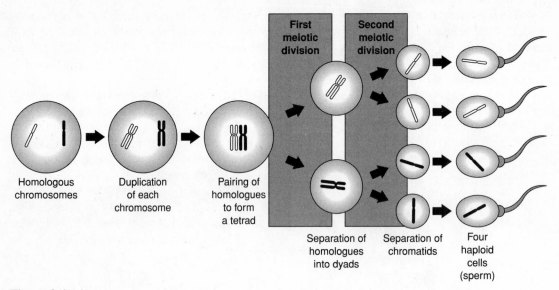

Figure 3.6 The process of meiosis depicted for only two chromosomes. Duplicated members of the pair of homologous chromosomes come to lie side by side in a four-stranded configuration (tetrad). Two successive divisions then result in formation of four haploid cells (sperm, in this case).

pair has duplicated itself, the result is a quadruple chromatid structure known as a *tetrad* (Greek, *tetras,* "a collection of four"). For the four chromatids to be distributed individually into four haploid cells, two divisions would be required. This is exactly what takes place. The two divisions accomplish (1) separation of the homologous chromosomes that had paired and (2) separation of the duplicated strands of each chromosome.

It would be instructive to amplify these remarks as we re-examine figure 3.6. During the first meiotic division, it may be seen that the tetrad is separated into two *dyads* (Greek, *dyas,* "a couple"). The first meiotic division is concerned with the separation of the original synaptic mates, or the original homologous members of the chromosomal pairs. The second meiotic division involves the separation of the chromatids, which had arisen earlier by duplication of the chromosomes. The overall outcome of meiosis is the formation of four haploid cells from one diploid cell. Each of the four haploid cells has half as many chromosomes as the original parent cell.

MENDEL AND THE CHROMOSOMAL THEORY OF HEREDITY

We shall now endeavor to relate Mendel's work to our present knowledge of chromosomes. Mendel, of course, was unaware that his hereditary factors were carried in the chromosomes. It was Walter Sutton in 1902 who first called attention to the parallelism in the behavior of genes and chromosomes. We have learned that chromosomes occur in pairs and that the members of a pair separate, or segregate, from each other during meiosis. Mendel's experiments reveal that the hereditary factors occur in pairs and that the members of each pair of factors separate from each other in the production of gametes. Evidently, then, the most reasonable inference is that the genes are located in chromosomes. Extensive studies have provided strong confirmation that the chromosomes do carry the genes. Indeed, chromosomes are the vehicles for transmitting the blueprints of traits to the offspring.

Figure 3.7 shows the separation of the pair of homologous chromosomes during meiosis of a heterozygote carrying the *A* and *a* alleles. The allele *A* occupies a particular site, or *locus,* in one of the chromosomes. The alternative allele, *a,* occurs at the identical locus in the other homologous chromosome. The alternative forms of a given gene, occupying a given locus, are termed *alleles.* Thus, *A* is an allele of *a,* or we may also say that *A* is allelic to *a.* As a summary statement, the term "allele" signifies a particular form of a gene, whereas the word "locus" means a position in a chromosome.

(Although we strive for precision of expressions, many writers, including geneticists, tend to be lax in the usage of terms. "Gene," "allele," and "locus" are frequently treated, improperly, as interchangeable terms.)

The heterozygote *(Aa)* produces two types of gametes, *A* and *a,* in equal numbers. It should be evident from figure 3.7 that the gametic ratio of 1*A*:1*a* is a consequence of the separation of the two homologous chromosomes during meiosis, one chromosome containing the *A* allele and its homologue carrying the *a* allele. The behavior of chromosomes also explains another feature of Mendel's results—the independent assortment of traits.

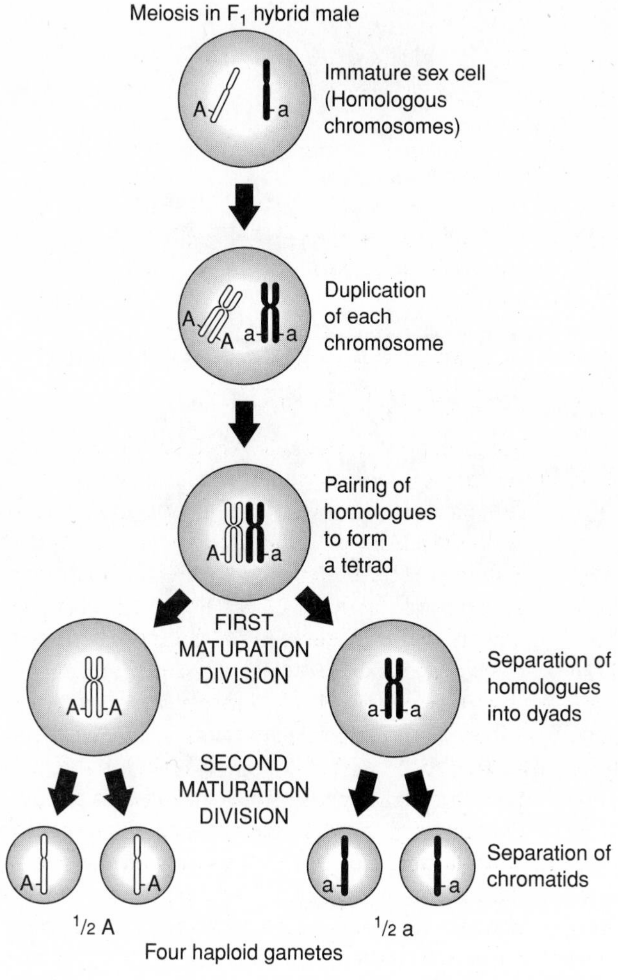

MENDEL'S PRINCIPLE OF INDEPENDENT ASSORTMENT

The fruit fly *(Drosophila melanogaster),* approximately one-fourth the size of a house fly, has long been a favorite subject for genetical research. Numerous varieties have been found, affecting all parts of the body. For example, the wings may be reduced in length to small vestiges, the result of action of a recessive *vestigial* allele. Two heterozygous normal-winged flies, each harboring the recessive allele for vestigial wings, can produce two types of offspring: normal-winged and vestigial-winged, in a ratio of three-to-one. Now suppose that these same parents have a second concealed recessive allele, namely, *ebony.* This recessive allele, when homozygous, modifies the normal gray color of the body to black, or ebony.

Two normal-winged, gray-bodied parents, each heterozygous *(NnGg),* can give rise to four types of offspring phenotypically, as seen in figure 3.8. Two of the phenotypes are like the original parents, and two are new combinations (normal-winged, ebony and vestigial-winged, gray). The four phenotypes appear in the

Figure 3.7 Chromosomal basis of monohybrid inheritance. When one chromosome of a pair carries a given gene (*A,* in this case) and its homologue carries the alternative form of the gene (*a*), then meiosis results in production of two distinct kinds of gametes in equal proportion, *A* and *a.*

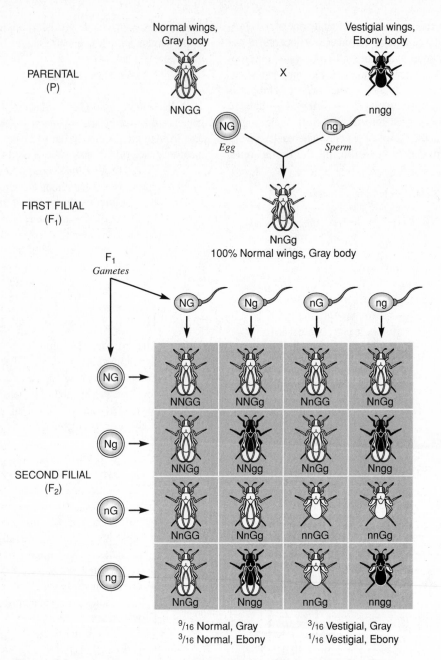

PARENTAL
(P)

Normal wings,
Gray body

X

Vestigial wings,
Ebony body

NNGG

nngg

NG   Egg

ng   Sperm

FIRST FILIAL
(F₁)

NnGg
100% Normal wings, Gray body

F₁
Gametes

NG   Ng   nG   ng

SECOND FILIAL
(F₂)

NG

NNGG   NNGg   NnGG   NnGg

Ng

NNGg   NNgg   NnGg   Nngg

nG

NnGG   NnGg   nnGG   nnGg

ng

NnGg   Nngg   nnGg   nngg

9/16 Normal, Gray   3/16 Vestigial, Gray
3/16 Normal, Ebony   1/16 Vestigial, Ebony

Figure 3.8 Mendel's law of independent assortment, as revealed in the inheritance of two pairs of characters in the fruit fly. A pure normal-winged fly with gray body mated with a vestigial-winged, ebony-bodied fly produces all normal-winged, gray-bodied flies in the first generation *(F₁)*. When these *F₁* flies are inbred, a second generation *(F₂)* is produced that displays a phenotypic ratio of 9:3:3:1.

classical Mendelian ratio of 9:3:3:1. For such a ratio to be obtained, each parent *(NnGg)* must have produced four kinds of gametes in equal proportions: *NG, Ng, nG,* and *ng.* Stated another way, the segregation of the members of one pair of alleles must have occurred independently of the segregation of the members of the other allelic pair during gamete formation. Thus, 50 percent of the gametes received *N,* and of these same gametes, half obtained *G* and the other half *g.* Accordingly, 25 percent of the gametes were *NG* and 25 percent were *Ng.* Likewise, 50 percent of the gametes carried *n,* of which half contained as well *G* and half *g* (or 25 percent *nG* and 25 percent *ng*). The four kinds of egg cells and the four kinds of sperm cells can unite in 16 possible ways, as shown graphically in the "checkerboard" in figure 3.8.

The above cross illustrates Mendel's *principle of independent assortment,* which states that one trait (one allelic pair) segregates independently of another trait (another allelic pair). The separation of the *Nn* pair of alleles and the simultaneous independent separation of the *Gg* pair of alleles occurs because the two kinds of genes are located in two different pairs of chromosomes. Figure 3.9 shows how the four types of gametes are produced with equal frequency in terms of the behavior of two pairs of chromosomes during meiosis. When the two pairs of alleles *N,n* and *G,g* are inserted at appropriate loci in the two pairs of chromosomes, it becomes apparent that four types of gametes, *NG, Ng, nG,* and *ng* are produced in

equal numbers. The parallelism in the behavior of genes and chromosomes is striking.

THREE OR MORE GENE PAIRS

When individuals differing in three characteristics are crossed, the situation is naturally more complex, but the principle of independent assortment still obtains. An individual heterozygous for three independently assorting pairs of alleles (for example,

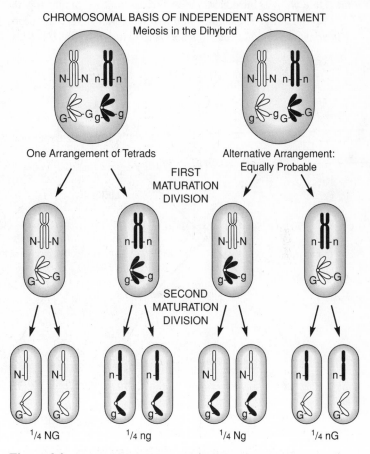

Figure 3.9 Chromosomal basis of dihybrid inheritance. The two pairs of genes, *N, n* and *G, g* are located in two pairs of chromosomes. When a dihybrid *(NnGg)* undergoes meiosis, there are two equally likely ways the tetrads may line up. The outcome is the formation of four kinds of gametes, in equal numbers *(NG, Ng, nG,* and *ng).*

Aa, *Bb*, and *Cc*) produces eight different types of gametes in equal numbers, as illustrated in figure 3.10. Random union among these eight kinds of gametes yields 64 equally possible combinations. Stated another way, Punnett's checkerboard method of analysis would have 64 (8 × 8) squares. An analysis of the offspring reveals that, among the 64 possible combinations, there are only eight visibly different phenotypes. You may wish to verify that the phenotypic ratio is 27:9:9:9:3:3:3:1.

The use of the Punnett square for three or more independently assorting genes is cumbersome. Here we will introduce one of the cardinal rules of probability: *The chance that two or more independent events will occur together is the prod-* *uct of their chances of occurring separately.* As an example, what proportion of the offspring of the cross *AaBbCc* × *AaBbCc* would be expected to have the genotype *AaBBcc*? The individual computations are as follows, treating each character (or event) independently:

1. The chance that an individual will be *Aa* = 1/2.
2. The chance that an individual will be *BB* = 1/4.
3. The chance that an individual will be *cc* = 1/4.

Assuming independence among these three pairs of alternatives, the chance of the three combining together *(AaBBcc)* is the product of their separate probabilities, as follows:

$$1/2 \times 1/4 \times 1/4 = 1/32.$$

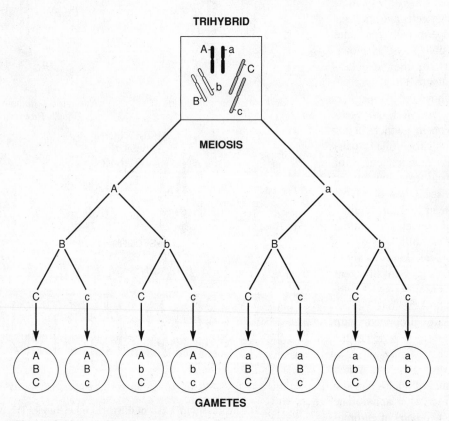

Figure 3.10 Gamete formation in an individual heterozygous for three pairs of alleles ("trihybrid"). The members of the three different pairs of alleles assort independently of each other when gametes are formed. The separation of the *Aa* pair, *Bb* pair, and the *Cc* pair, respectively, can be treated as independent events. The outcome is eight genetically different gametes.

SIGNIFICANCE OF GENETIC RECOMBINATION

The Mendelian principles permit a genuine appreciation of the source of heritable variation in natural populations of organisms. An impressively large array of different kinds of individuals can arise from the simple processes of segregation and recombination of independently assorting genes. Figure 3.11 considers a few of the independently assorting traits in the fruit fly. It is seen that the number of visibly different classes (phenotypes) of offspring is doubled by each additional heterozygous pair of independently assorting alleles. Each additional heterozygous allelic pair multiplies the number of visibly different combinations by a factor of two. Thus, if the parents are each heterozygous for ten pairs of genes (when dominance is complete), the number of different phenotypes among the progeny becomes 2^{10}, or 1,024.

Other mathematical regularities are evident as we increase the number of heterozygous pairs

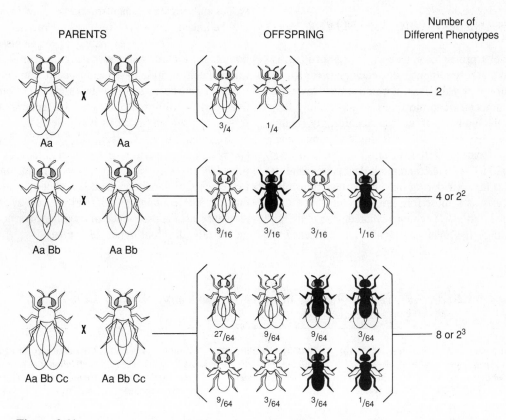

Figure 3.11 Inheritance of traits in the fruit fly, when the parents are each heterozygous for one *(Aa)* independently assorting trait and for two *(AaBb)* and three *(AaBbCc)* independently assorting traits. The allele for long wings *(A)* is dominant over the allele for vestigial wings *(a);* the allele for gray body *(B)* is dominant over the allele for ebony *(b);* and the allele for normal eyes *(C)* is dominant over the allele producing the eyeless condition *(c).* The number of visibly different classes (phenotypes) of offspring increases progressively with each additional pair of traits. The generalized formula for determining the number of different phenotypes among the offspring (when dominance is complete) is 2^n, where *n* stands for the number of different heterozygous pairs of alleles.

(table 3.2). As the number of heterozygous traits increases, the chance of recovering one of the parental types becomes progressively less. Thus, when a single heterozygous pair *(Aa)* is involved, 1 in 4 will resemble one of the original parents in both appearance and genotype (*AA* or *aa*). When four heterozygous traits are involved, only 1 in 256 will be genotypically like either of the original parents. Evidently, no single genetic constitution is ever likely to be exactly duplicated in a person (save in identical twins). Each individual is genetically unique.

LINKAGE AND CROSSING OVER

Mendel's principle of independent assortment is based on the assumption that each gene for a trait is carried in a different chromosome. However, the number of chromosomes is appreciably less than the number of genes. Accordingly, many genes must be carried in a single chromosome. Genes located together in any one chromosome are said to be *linked,* and the linked genes tend to remain together during their hereditary transmission. It was a most fortuitous circumstance that Mendel did not encounter linked genes in his breeding experiments. If Mendel had happened to

work on two linked genes, he certainly would have come upon some puzzling ratios among the offspring.

The principle of linkage was firmly established in the early 1900s by T.H. Morgan and his students through studies on the fruit fly. The fruit fly *(Drosophila melanogaster)* has only four pairs of chromosomes. Several hundred different genes have been identified in the fruit fly, and all fall into four linkage groups corresponding to the four pairs of chromosomes. Humans have 23 linkage groups.

The effect of linkage is to decrease variability, but an intricate mechanism exists for rearranging the linked genes. As a result of the mechanism of *crossing over*—the exchange of segments of homologous chromosomes during meiosis—the original combination of genes in a particular chromosome can be broken up. In essence, crossing over permits different alleles of one gene to be tested in new combinations with alleles of other genes. In an illustrative example (fig. 3.12), if the two pairs of genes had been so closely linked as to preclude any crossing over between them, only two classes of gametes would have been formed; one gamete containing *AB,* and the other, *ab.* However, as a consequence of crossing over, four genetically different types of gametes are pro-

TABLE 3.2	Characteristics of Crosses Involving Various Pairs of Independently Assorting Alleles			
Number of heterozygous pairs involved in the cross	Number of different types of gametes produced by the heterozygote	Number of visibly different progeny (that is, phenotypes)*	Number of genotypically different combinations (that is, genotypes)	Chance of recovering an individual that is either homozygous dominant or recessive for all traits
1	2	2	3	1 in 4
2	4	4	9	1 in 16
3	8	8	27	1 in 64
4	16	16	81	1 in 256
10	1,024	1,024	59,049	1 in 1,048,576
n	2^n	2^n	3^n	1 in 4^n

*Assumes complete dominance in each pair.

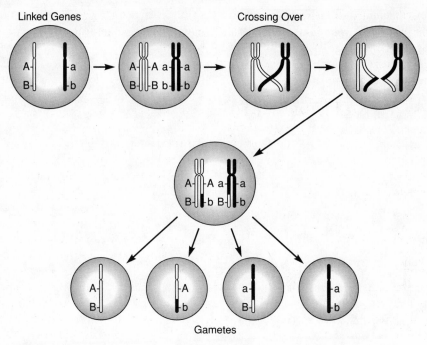

Linked Genes   Crossing Over

Gametes

Figure 3.12 **Crossing over the chromosomal material** during the tetrad stage of meiosis. If the two pairs of genes were not linked, the hybrid individual *(AaBb)* would produce four genetically different gametes in equal proportions *(AB, Ab, aB,* and *ab)*. A hybrid containing the two pairs of linked genes also produces four genetically different gametes as a consequence of crossing over, but the new combinations of linked genes, or recombination types *(Ab* and *aB)*, arise less frequently than the original combination types *(AB* and *ab)*. The magnitude of deviation from a 1:1:1:1 ratio depends on the frequency with which crossing over occurs between the linked genes.

duced. Thus, crossing over provides a mechanism for increasing variability—that is, rearranging genes in new combinations.

The reshuffling of genes by independent assortment, coupled with the recombination of genes through crossing over, may be compared, on a very modest scale, to the shuffling and dealing of playing cards. One pack of 52 cards can yield a large variety of hands. And, just as in poker a full house is far superior to three of a kind, so in organisms certain combinations of genes confer a greater reproductive advantage to their bearer than other combinations. Natural selection favors the more reproductive genotypes.

CHAPTER

4

MUTATION

In the opening pages of *The Origin of Species,* Charles Darwin declared, "Any variation which is not inherited is unimportant for us." This and other incisive statements show that Darwin understood the importance of inherited variability. He was, however, eternally assailed by doubts as to how new varieties initially arise. Today we know that the ultimate source of genetic variation is *mutation.*

Each gene of an organism may assume a variety of forms. A normal gene may change to another form and produce an effect on a trait different from that of the original gene. An inheritable change in the structure of the gene is known as a mutation. Variant traits, such as albinism in humans and vestigial wings in the fruit fly, are traceable to the action of altered, or mutant, genes. *All differences in the genes of organisms have their origin in mutation.*

CAUSES OF MUTATION

New mutations arise from time to time, and the same mutations may occur repeatedly. It is often difficult to distinguish between new mutations and old ones that occurred previously and were carried concealed in ancestors. A recessive mutant gene may remain masked by its normal dominant allele for many generations and reveal itself for the first time only when two heterozygous carriers of the same mutant gene happen to mate.

Each gene runs the risk of changing to an alternative form. The causes of naturally occurring, or *spontaneous,* mutations are largely unknown. The environment contains a background of inescapable radiation from radioactive elements, cosmic rays and gamma rays. It is generally conceded that the amount of background radiation is too low to account for all spontaneous mutations. In other words, only a small fraction of spontaneous mutations can be attributed to background radiation. We may say that spontaneous variations are random and unpredictable.

In 1927, the late Nobel laureate Hermann J. Muller of Indiana University announced that genes are highly susceptible to the action of x rays. By irradiating fruit flies with x rays, he demonstrated that the process of mutation is enormously speeded up. The production of mutations is directly proportional to the total dosage of x rays. Stated another way, the greater the dosage, the greater the mutation rate. In this linear, or straight-line, relationship, the mutation rate increases proportionately about 3 percent for each 1,000 units (or *roentgens*) of x rays.

It has long been held that the mutagenic effect is the same whether the dose is given in a short time or spread over a long period. In other words, low intensities of x rays over long periods of time produce as many mutations as the same dose administered in high intensities in a short period of time. Recent experiments on mice have cast some doubt on this view, for it has been shown (at least in mice) that the mutagenic effect of a single exposure to the germ cells is greater than the effect of the same exposure administered as several smaller doses separated by intervals of time. Nevertheless, there does not appear to be a critical, or *threshold,* dose of roentgens below which there is no effect. In essence, no dose is so low (or "safe") as to carry no risk of inducing a mutation. Modern workers stress that any amount of radiation, no matter how little, can cause a mutation.

At the time of Muller's discovery, no one conceived that within a generation the entire population of humans would be exposed to a significant increase of high-energy radiation as a consequence of the creation of the atomic bomb. The additional amount of high-energy radiation already produced by fallout from atomic explosions has undoubtedly increased the mutation rate. Most of the radiation-induced mutations are recessive and most of them are deleterious.

In discussing the hazards of x rays and other ionizing radiations, we must be careful to distinguish between *somatic damage* and *genetic damage.* Injury to the body cells of the exposed individuals themselves constitutes somatic damage. On the other hand, impairment of the genetic apparatus of the sex cells represents genetic damage. Typically, the genetic alterations do not manifest themselves in the immediate persons but present a risk for their descendants in the next or succeeding generations.

There are documented records of the *somatic* consequences of exposure to the 1945 atomic blasts in Hiroshima and Nagasaki. Among 161 children born of women who were exposed to the atomic bomb while pregnant, 29 were microcephalic (head size considerably below normal) and 11 of these 29 were mentally retarded. As might have been expected, the detrimental effects were most pronounced among the infants of women who were in the early stage of pregnancy (less than 15 weeks' gestation) at the time of exposure. Moreover, most of the women who gave birth to deformed infants were less than 1.3 miles from the center (hypocenter) of the explosion. The adverse effects on the fetus diminished in frequency and severity as the distance from the hypocenter increased.

Analyses have also revealed that survivors of the atomic blasts have developed leukemia in rough proportion to the dosage of ionizing radiation. From 0.02 to 1.00 percent of the survivors within the range of radiation developed leukemia in the decade between 1950 and 1960, with the greatest percentage of cases present in individuals nearest the hypocenter. The data substantiate the association found in other studies between whole-body exposure to radiation at high dose levels and the incidence of leukemia.

Estimates of the magnitude of possible *genetic* damages have been of uncertain significance. The statistical methods used are too insensitive to detect the occurrence of radiation-induced mutations. There were no more gene-determined defects in infants born to women who were pregnant at the time of the atomic blasts at Hiroshima and Nagasaki than there were in populations that were not exposed. However, it must be stressed that most of the possible recessive mutations induced would not be expressed for several generations.

In an effort to measure immediate damaging genetic effects, a possible shift in the sex ratios among children of exposed parents was sought. Theoretically, if fatal, or lethal, mutations are induced on the X chromosome, they will be immediately expressed in the sons of the exposed women (fig. 4.1). Deaths of male fetuses

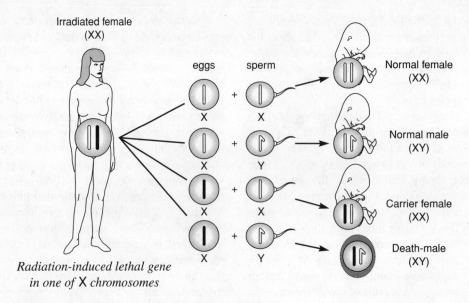

Figure 4.1 Death of male fetuses resulting from radiation-induced lethal mutations in one of the X chromosomes of the mother. Theoretically, the sex ratio (the proportion of males to females) among children conceived after exposure of the expectant mother to radiation should be diminished.

prior to birth can be indirectly gauged by a decrease in the number of males born to irradiated mothers. An early, limited study on the offspring of Japanese mothers who received a heavy dose of radiation revealed a slight reduction of male births, but a later study, based on a larger sample, showed no alteration of the sex ratio.

Although the actual data available have failed to reveal unequivocal genetic effects of radiation, it would be a patent fallacy to conclude that atomic radiation has had no mutagenic effect. American and British geneticists estimate that each person currently receives a total dose of 7.8 roentgens to the reproductive cells during the first 30 years of life (table 4.1). Of this amount, 3.1 roentgens are derived from natural background radiation, 4.6 roentgens from various medical uses of ionizing radiation, and an additional 0.1 roentgen from the testing of nuclear weapons. The additional amount of radiation (0.1 roentgen) received from nuclear fallout may seem trivial.

TABLE 4.1	Average Exposure to Radiation of the Gonads (In Roentgen units, R)
Sources of Radiation	
1. Background	3.1 R per 30 years
External	
Cosmic rays—0.84	
Terrestrial—1.41	
Atmospheric—0.16	
Internal	
Natural radioactive atoms (^{40}K, ^{14}C, etc.)—0.69	
2. Medical uses of radiation	4.6 R
3. Fallout from exposures	<u>0.1 R</u>
	7.8 R per 30 years

But, the exposure of our population to 0.1 roentgen is calculated to induce sufficient mutations to result in 3,750 defective offspring among 100 million births.

CHEMICAL NATURE OF THE GENE

During the early decades of this century, the picture of genes that had emerged was that of discrete entities, strung together along the chromosome, like individual beads on a string. This portrayal carried with it the implication that genes were not pure abstractions, but actual particles of living matter. Yet, until the 1950s the chemical nature of the gene remained enigmatic. Biochemists wondered whether there existed an elaborate organic molecule with a simple string-like form.

One of the masterly feats of modern science has been the elucidation of the chemical nature of the gene. The transmission of traits from parents to offspring depends on the transfer of a specific giant molecule that carries a coded blueprint in its molecular structure. This complex molecule, the basic chemical component of the chromosome, is *deoxyribonucleic acid* (DNA). The information carried in the DNA molecule can be divided into a number of separable units, which we now recognize as the genes. Stated simply, chromosomes are primarily long strands of DNA, and genes are coded sequences of the DNA molecule. Genes specify the structure of individual proteins. The diverse genes in a single fertilized human egg can code an estimated 50,000 to 100,000 proteins. The once theoretical gene has finally been placed on a material (chemical) level.

The two scientists who worked together in the early 1950s to propose a configuration for the DNA molecule were Francis H.C. Crick, a biophysicist at Cambridge University, and James D. Watson, an American student of virology who was then studying chemistry at Cambridge on a postdoctoral fellowship. With the invaluable aid of x-ray diffraction pictures of DNA crystals prepared by Maurice H. F. Wilkins and Rosalind Franklin at King's College in London, Watson and Crick built an inspired model in metal of DNA's configuration. This achievement won Watson, Crick, and Wilkins the coveted Nobel Prize for physiology or medicine in 1962. The other deserving scientist, Rosalind Franklin, unfortunately died of cancer before the Nobel Prize (awarded only to living persons) was announced.

The remarkable feature of DNA is its simplicity of design. The double-stranded DNA molecule is shaped like a twisted ladder (fig. 4.2). The two parallel supports of the ladder are made up of alternating units of sugar (deoxyribose) and phosphate molecules, while the cross-links, or rungs, are composed of specific nitrogen-containing ring compounds, or nitrogenous bases. Each rung of the ladder consists of one pair of nitrogenous bases, held together by specific hydrogen bonds. Hydrogen bonds are weak bonds. However, the sum total of the hydrogen bonds between the two strands assures that the two strands of the double helix are firmly associated with each other under conditions commonly found in living cells.

There are two classes of nitrogenous bases, the larger, two-ring purines and the smaller, one-ring pyrimidines. In making up the model, Watson and Crick found that the nitrogenous bases could be accommodated best if the double-ring purines (adenine or guanine) lay opposite the single-ring pyrimidines (thymine or cytosine). The arrangement of the bases is not haphazard: adenine (A) in one chain is typically coupled to thymine (T) in the other chain, and guanine (G) is preferentially linked with cytosine (C). Along one chain, any sequence of the bases is possible, but if the sequence along one chain is given, then the sequence along the other is automatically determined, because of the precise pairing rule (A=T and G=C). The combination of one purine and one pyrimidine to make up each cross connection is conveniently called a *base pair.*

The DNA molecule has a structure that is sufficiently versatile to account for the great variety of different genes. The four bases (A, C, G, and T) may be thought of as a four-letter alphabet or code. The uniqueness of a given gene would then relate to the specific order or arrangement of the bases, just as words in our language differ according to the sequence of letters of the alphabet or as a telegraphic message becomes comprehensible by the varied combinations of dots and dashes. The number of ways in which the nitrogenous bases can be arranged in the DNA molecule is exceedingly large. The human fertilized egg contains 46 chromosomes with about 6.6 billion base pairs of

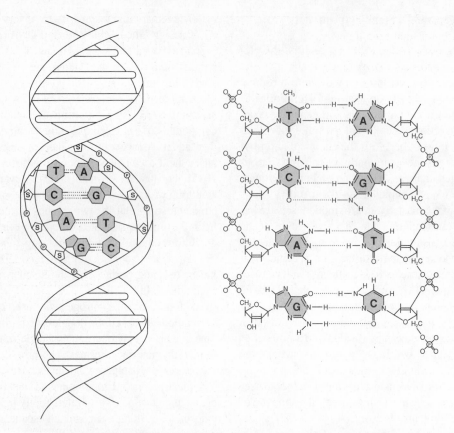

Figure 4.2 Watson-Crick double-stranded helix configuration of deoxyribonucleic acid (DNA). The backbone of each twisted strand consists of alternating sugar (S) and phosphate (P) residues. Enlarged view on right shows that the larger, two-ring purines (adenine or guanine) lay opposite the smaller, one-ring pyrimidines (cytosine or thymine). The nitrogenous bases are held together by hydrogen bonds, three between cytosine (C) and guanine (G), and two between adenine (A) and thymine (T).

DNA. As a generalization, a single gene is a linear sequence of approximately 1,500 base pairs.

SELF-COPYING (REPLICATION) OF DNA MOLECULE

One fundamental property that has long been ascribed to the gene is its ability to make an exact copy of itself. If the gene is really a linear sequence of base pairs, then the Watson-Crick model of DNA must be able to account for the precise reproduction of the sequences of the bases. The replication machinery is quite complicated, but conceptually the self-copying of the double helix can be visualized with little difficulty.

The DNA molecule consists of two chains, each of which is the complement of the other. Accordingly, each single chain can serve as a mold, or *template,* to guide the formation of a new companion chain (fig. 4.3). The two parallel chains separate, breaking the hydrogen bonds that hold together the paired bases. Each chain then attracts new base units from among the supply of free units always present in the cell. As disassociation of the strands takes place, each separated lengthwise portion of the chain can begin to form a portion of the new chain. Eventually, the entire original double helix has produced two exact replicas of itself. If the original two chains are designated *A* and *B,* the old *A* will direct the formation of a new *B* and the old *B* will guide the production of a new *A.* Where one *AB* molecule existed previously, two *AB* molecules, exactly like the original, exist afterward.

Although the replication of a DNA strand is remarkably accurate, errors do occur in the positioning, or insertion, of the nitrogenous bases in a growing chain. The cell has several repair ("proofreading") mechanisms by which incorrectly inserted bases, or otherwise altered bases, are replaced by the appropriate bases. Occasionally, however, the cell's DNA repair program fails, and an accidental mistake endures. Such a sustained error qualifies as a *base substitution mutation.* Indeed, the vast majority of single base changes occur during DNA replication. An uncorrected base change can replicate as faithfully as the formerly normal base. It is estimated that a base substitution mutation occurs once in every one billion (1×10^{-9}) bases added during the replication process. A mutational event is, accordingly, rare. Yet, if the cell's DNA repair

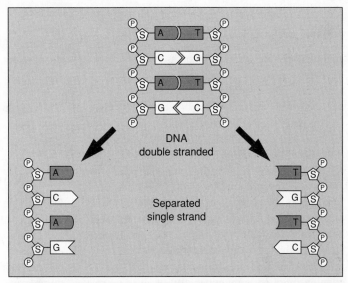

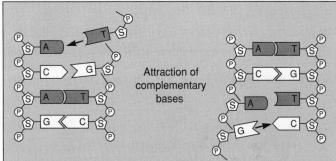

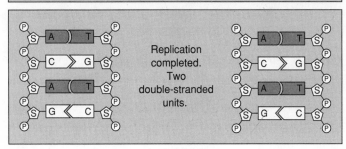

Figure 4.3 Replication of DNA. The parental strands separate, each serving as a guide or template for the synthesis of a new daughter strand. Each new DNA molecule contains one parental and one daughter strand.

mechanism were ever to become infallible, then evolution, as we know it, would cease for the lack of mutations.

MOLECULAR MECHANISM OF SPONTANEOUS MUTATION

A mutation may be thought of as a spontaneous mishap to one of the paired bases in the DNA molecule. A single substitution of a C-G pair for a T-A may be sufficient to alter the character of the gene. One possible way in which such a substitution might occur is shown in figure 4.4. It has been established that each of the bases in DNA, as the result of random thermal motions of their atoms, can exist in several rare forms known as *tautomers.* The tautomeric alternatives differ in the positions at which certain atoms are attached in the molecule. The significance of the tautomeric alternatives lies in the changed pairing qualities they impart. Whereas, for example, the normal form of adenine pairs with thymine, one of its rare tautomeric forms actually attracts and couples with cytosine. Such a tautomeric shift in the adenine molecule may thus make possible a new purine-pyrimidine base pair arrangement in the DNA molecule. As seen in figure 4.4, an adenine

(labeled A′) has undergone a tautomeric shift and attracts the wrong partner (cytosine instead of thymine). At the next replication of the strand, the misattracted cytosine acts normally to join with guanine. Hence, a C-G pair is established where a T-A pair was formally located. At this point in the DNA molecule, the gene is modified and may be expected to produce a mutant effect.

LANGUAGE OF LIFE

Cellular proteins are key components of living matter. Proteins are composed of 20 basic building blocks, the *amino acids,* arranged in long chains, called *polypeptides.* Each polypeptide chain may be several hundred amino acid units in length. The number of possible arrangements of the different amino acids in a given protein is unbelievably large. In a polypeptide chain made up of 500 amino acid units, the number of possible patterns (given 20 different amino acids) can be expressed by the number 1 followed by 1,100 zeros. How,

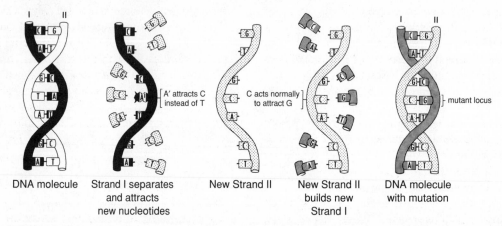

DNA molecule   Strand I separates   New Strand II   New Strand II   DNA molecule
                 and attracts   builds new   with mutation
                 new nucleotides   Strand I

A′ attracts C instead of T

C acts normally to attract G

mutant locus

Figure 4.4 Mutation at the molecular level. When cell division occurs, the two twisted strands separate and each strand attracts unbound nucleotides (containing nitrogenous bases) to rebuild the DNA molecule. This particular illustration shows events through two successive cell divisions, commencing with strand I in the original DNA molecule. During the first division, an adenine radical (A) undergoes a chemical change to A′, which misattracts cytosine (C) instead of thymine (T). At the next division, cytosine (C) attracts its customary partner, guanine (G). The net result is that the granddaughter DNA molecule contains a C-G pair where a T-A pair was formerly located. This highly localized change in one of the pairs of nitrogenous bases constitutes a gene mutation.

then, does a cell form the particular amino acid patterns it requires out of the colossal number of patterns that are possible?

We may presume that the sequence of bases of the DNA molecule is in some way the master pattern, or code, for the sequence of the amino acids composing polypeptides. In 1954, the British physicist George Gamow suggested that each amino acid is dictated by one sequence of three bases in the DNA molecule. As an example, the sequence cytosine-thymine-thymine (CTT) in the DNA molecule might designate that a particular amino acid (glutamic acid) be incorporated in the formation of a protein molecule, such as hemoglobin (fig. 4.5). Thus, the DNA code is to be found in *triplets*—that is, three bases taken together code one amino acid. It should be noted that only one of the two strands of the DNA molecule serves as the genetic code. (More precisely, one strand is consistent for a given gene, but the strand varies from one gene to another.) Biochemical chaos would result if both complementary strands of DNA were to encode information.

The importance of the genetic code cannot be overstated. Let us assume the following sequence of bases in one of the strands of the DNA molecule: ... CGT ATC GTA AGC ..., and that these triplets specify four amino acids, designated **R, I, P,** and **E.** In other words, the following code exists: CGT=**R,** ATC=**I,** GTA=**P,** and AGC=**E.** This section of the DNA molecule thus specifies **RIPE,** and the message to make **RIPE** is passed on by this particular sequence from the nucleus to the cytoplasm of the cell. The message continues to flow out in the living cell, and **RIPE** will be made, copy after copy.

But what if a chance mishap occurs in one of the triplets? Let us say that T in the second triplet is substituted for A, so that the triplet reads TTC instead of ATC, which would specify an **O** instead of **I.** The word would now be **ROPE** instead of **RIPE.** This makes quite a difference, especially if the word is continually printed incorrectly throughout a novel. Just as a misprinted word can alter or destroy the meaning of a sentence, so an altered

protein in the body fails to express its intended purpose. Sometimes the error is not tragic, but often the organism is debilitated by the misprint.

As a striking example, we may consider the case of individuals afflicted with sickle-cell anemia. The full story of sickle-cell anemia is narrated in chapter 8. For the moment, it suffices to say that the hemoglobin molecule in sickle-cell anemic patients is biochemically abnormal. The sole difference in chemical composition between normal and sickle-cell hemoglobin is the substitution of only one amino acid unit among several hundred. Specifically, at one site in a particular polypeptide chain of the sickle-cell hemoglobin molecule, the amino acid *glutamic acid* has been replaced by *valine.* The detrimental effect of sickle-cell anemia is thus traceable to an exceedingly slight alteration in the structure of the protein molecule. This, in turn, is associated with a simple base-pair switch, or mutation, in the DNA molecule in the chromosome (fig. 4.5).

The DNA molecule, like a tape recording, carries specific messages for the synthesis of a wide variety of proteins. We shall see presently that DNA does not directly form protein but works in an intriguing way through a secondary form of nucleic acid, *ribonucleic acid,* or RNA. At this point, it is important to recognize that a gene is a coded sequence in the DNA molecule. From a functional point of view, we may say that a gene is *a section of the DNA molecule (about 1,500 base pairs) involved in the determination of the amino acid sequence of a single polypeptide chain of a protein.*

TRANSCRIPTION OF DNA

It has been said that hemoglobin served as the Rosetta stone in revealing the hereditary language. It was the hemoglobin molecule that provided the first direct proof of the relationship of the gene to protein synthesis. Proteins are not synthesized directly on the DNA template. The genetic information (genetic code) of DNA is first transcribed to ribonucleic acid (RNA). The RNA molecule

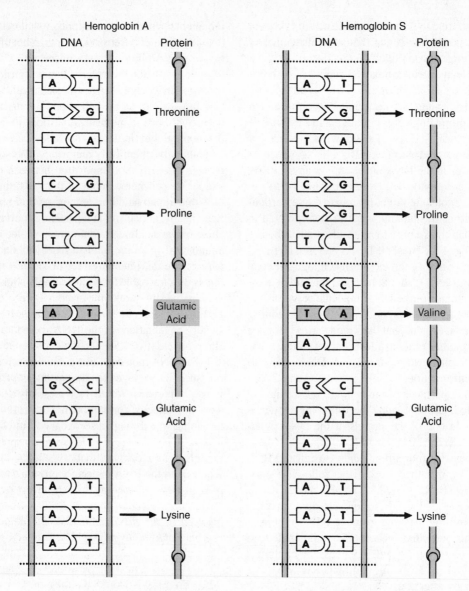

Figure 4.5 The abnormal hemoglobin that occurs in sickle-cell anemia (hemoglobin S) is the consequence of an alteration of a single base of the triplet of DNA that specifies a particular amino acid (glutamic acid) in normal hemoglobin (hemoglobin A). A simple base-pair switch, or mutation, in the DNA molecule results in the replacement of glutamic acid in hemoglobin A by another amino acid, valine, in hemoglobin S. (Based on studies by Vernon M. Ingram.)

resembles the DNA molecule in structure except in three important respects. The pyrimidine base, *uracil,* is found in RNA, which replaces the thymine that is characteristic of DNA. Secondly, the sugar in RNA is *ribose,* which contains one more oxygen atom than does the deoxyribose sugar. Thirdly, RNA has only a *single* strand instead of two.

As seen in figure 4.6, one of the two strands of DNA (always the same one for a given gene) forms a complementary strand of RNA. The same rules of pairing hold as in replicating a copy of DNA, except that adenine attracts uracil instead of thymine. This RNA strand, which is responsible for carrying DNA's instructions out into the cytoplasm, is appropriately termed *messenger RNA,* or mRNA. The amino acids of a polypeptide chain are specified by the messenger RNA. In a large sense, then, the genetic code applies to RNA rather than to DNA itself.

Scientists have broken the genetic code in its main elements. Each three-letter unit of the messenger RNA is called a *codon.* Each of the twenty main amino acids, ranging from alanine to valine, is specified by at least one codon. In fact, most amino acids have more than one codon. Serine, for example, has six codons; glycine has four; and lysine has two. Only methionine and tryptophan have one each. Because most amino acids are specified by more than one codon, the code is claimed to be degenerate. However, this kind of degeneracy is welcome since it ensures that the code works.

Details of the code are shown in table 4.2. This table of the genetic code represents for the biologist what the periodic table of the elements represents for the chemist. It may be noted that three of the codons (UAA, UAG, and UGA) do not code for any of the 20 amino acids. They were originally, perhaps facetiously, described as "nonsense codons," but they are now known to function as *terminating codons* (or *stop codons*). They serve to signal the termination of a polypeptide chain.

One of the most impressive findings is that the genetic code is essentially universal. The same codon calls forth the same amino acid in organisms as

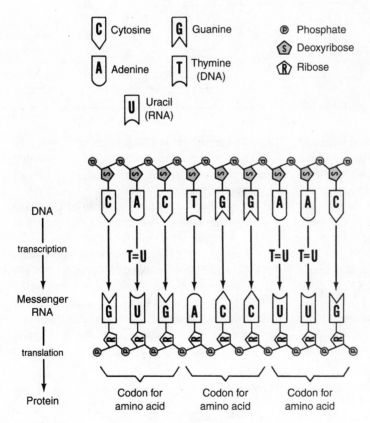

Figure 4.6 The language of life. The DNA molecule forms a single strand of messenger RNA that carries DNA's instructions out into the cell. The four bases of DNA (A, T, C, and G) are responsible for the 3-letter code words, or codons, of messenger RNA. Each of the twenty amino acids that make up the variety of body proteins is specified by at least one codon.

TABLE 4.2	The Three-Letter Codons of RNA and the Amino Acids Specified by the Codons

AAU AAC	} Asparagine	CAU CAC	} Histidine	GQU GAC	} Aspartic acid	UAU UAC	} Tyrosine
AAA AAG	} Lysine	CAA CAG	} Glutamine	GAA GAG	} Glutamic acid	UAA UAG	} (Terminator)*
ACU ACC ACA ACG	} Threonine	CCU CCC CCA CCG	} Proline	GCU GCC GCA GCG	} Alanine	UCU UCC UCA UCG	} Serine
AGU AGC	} Serine	CGU CGC	} Arginine	GGU GGC	} Glycine	UGU UGC	} Cysteine
AGA AGG	} Arginine	CGA CGG		GGA GGG		UGA UGG	(Terminator)* Tryptophan
AUU AUC AUA	} Isoleucine	CUU CUC CUA	} Leucine	GUU GUC GUA	} Valine	UUU UUC	} Phenylalanine
AUG	Methionine	CUG		GUG		UUA UUG	} Leucine

* Terminating codons signal the end of the formation of a polypeptide chain.

widely separate phylogenetically as a bacterium *(E.coli)*, a flowering plant (wheat), an amphibian (the South African clawed frog), and mammals (guinea pig, mouse, and human). The essential universality of the genetic code supports the view that the code had its origin at least by the time bacteria evolved, 3 billion years ago. The genetic code may be said to be the oldest of languages.

The flow of genetic information is sketched in summary form in figure 4.7. The DNA molecule has the capacity to replicate as well as transcribe. The *transcription* of DNA into messenger RNA is the means by which genetic information is communicated from the nucleus to the cytoplasm. The transfer of information from messenger RNA to proteins is termed *translation.*

We may now take a closer look at sickle-cell hemoglobin in light of our knowledge of transcription and translation. This abnormal hemoglobin is caused by a mutation in one gene that codes a poypeptide chain, the so-called β-globin chain that is 146 amino acid residues in length. As we learned earlier, the mutation alters only one amino

acid unit in the entire chain, changing a glutamic acid present in normal hemoglobin into another amino acid, valine. The substitution of valine for glutamic acid can be accounted for by an alteration of a single base change of the triplet of DNA that specifies the amino acid, as depicted in figure 4.8. If the DNA triplet responsible for transcription in normal hemoglobin is CTT, then a single base change to CAT would alter the codon of messenger RNA to GUA. Accordingly, the amino acid specified by the codon GUA would be valine. The replacement of the charged glutamic acid by the neutral valine causes a change in the conformation of the hemoglobin molecule. This, in turn, causes a distortion in the morphology ("sickling") of the red blood cell.

FREQUENCY OF MUTATIONS IN HUMAN DISORDERS

Most mutations present in an individual are inherited from the previous generation. The rate at which new disease-causing mutations arise in

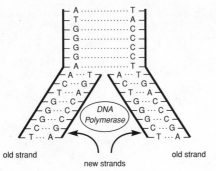

Replication

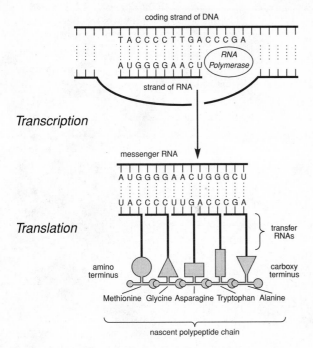

Transcription

Translation

Figure 4.7 Stepwise flow of information from DNA to RNA to protein. DNA not only copies itself by the act of *replication* with the aid of an enzyme (DNA polymerase), but expresses its information by generating a single-stranded RNA whose nucleotide sequence is coded by one of the two strands of the DNA duplex. The *transcription* of DNA into RNA is accomplished with the aid of an RNA polymerase. In the act of *translation,* the nucleotide sequence of the RNA transcript (messenger RNA) directs the sequence of amino acids comprising a protein. Transcription of DNA into RNA and translation of messenger RNA into protein are compartmentalized activities, the former event occurring in the nucleus and the latter in the cytoplasm.

humans is very low. In our present state of knowledge, the mutation rate for any single human gene must be considered as a rough estimate. It is likely that all human offspring contain at least one newly mutated gene capable of having an adverse effect. The consensus is that the rate of mutation per gene in the human ranges from one in 100,000 (1×10^{-5}) gametes per generation to one in 1,000,000 (1×10^{-6}) gametes per generation. Stated another way, any given gene mutates, with a clinically adverse effect, approximately once in every 100,000 to 1,000,000 sperm cells or egg cells produced in a generation. A large-sized gene with many bases has a greater likelihood of experiencing a mutational change than a small-sized gene, just as the likelihood of a misspelling in a manuscript is greater in a long sentence than in a short sentence.

It may be noted that the mutation rate for a recognizable human disorder is higher than the estimate, stated earlier in this chapter, of the frequency of uncorrected base changes in the DNA molecule. This reflects the fact that not all single base changes (mutations) within a given gene have adverse effects or cause overt disease. Indeed, several hundred different base changes are known to occur in the genes that code for hemoglobin, but many of them are inconsequential to the properties or activity of the hemoglobin molecule and are, accordingly, innocuous. There are, however, approximately 400 hemoglobin variants that have been identified as having a clinical impact. Sickle-cell anemia, earlier discussed, represents a prominent example of a single base change that results in a marked impairment of the hemoglobin molecule.

The mutation responsible for *achondroplasia* is notable in that it constitutes one of the most mutable single base changes in the human genome. Achondroplasia is transmitted as an autosomal dominant trait. Affected individuals are small and disproportionate,

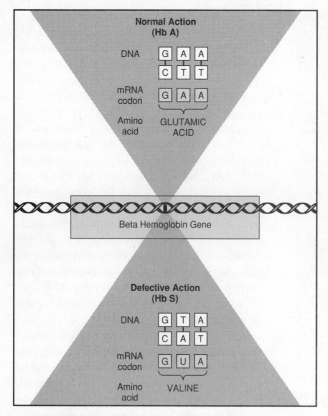

DNA
G A A
C T T

mRNA codon
G A A

Amino acid   GLUTAMIC ACID

Normal Action (Hb A)

Beta Hemoglobin Gene

Defective Action (Hb S)

DNA
G T A
C A T

mRNA codon
G U A

Amino acid   VALINE

Figure 4.8 **Derivation of abnormal hemoglobin S from** normal hemoglobin A by a single nucleotide change in the triplet of DNA bases that normally codes for glutamic acid.

with relatively long trunks and short arms and legs (fig. 4.9). Achondroplasia is a *congenital defect,* a defect present at birth. With an estimated prevalence of 1 per 40,000 live births, it is one of the more common Mendelian disorders.

Many achondroplastic dwarfs are stillborn or die in infancy; those surviving to adulthood produce fewer offspring than normal. About 80 percent of the children born with this condition in one generation will not replace themselves in the next generation. Given the high mortality and low fecundity, the frequency of the disorder would steadily decrease from one generation to the next, if it were not for the high mutability of the

gene involved. In fact, more than 90 percent of affected individuals have newly arisen mutations. Interestingly, an increased paternal age at the time of conception has been observed. That is to say, *de novo* mutations occur more often in older fathers than in older mothers. Indeed, in several other dominantly inherited disorders, normal parents more often produce an affected child when the father is older at the time of conception.

Every mutation thus far detected in achondroplastic dwarfs has been exactly at the same nitrogenous base in the gene, in which a guanine has been replaced by adenine. The outcome has been the substitution of the amino acid, glycine, by the amino acid, arginine, in one site in the protein coded by the aberrant gene. Achondroplasia is like sickle-cell anemia in that it is caused by a single, specific amino acid substitution. However, this dominant disorder differs markedly from sickle-cell anemia in that new mutations account for 90 percent of the cases. The mutation rate for the achondroplastic gene (more precisely, the specific guanine site) has been placed as high as 4.2×10^{-5}, or about 4 mutations in every 100,000 gametes produced in a generation. This particular guanine site is said to be a mutational "hot spot."

EVOLUTIONARY CONSEQUENCES OF MUTATIONS

By far, most of the gene mutations arising today in organisms are not likely to be beneficial. Existing populations of organisms are products of a long evolutionary past. The genes that are now normal to the members of a population represent the most favorable mutations selectively accumulated over eons of time. The chance that a new mutant gene

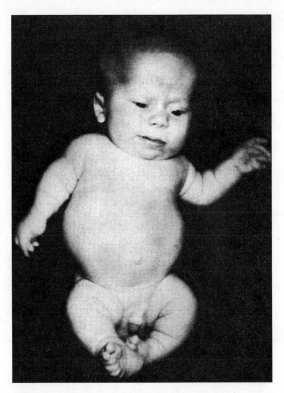

Figure 4.9 Achondroplastic dwarfism, a dominant genetic disorder in which the affected infant has inherited abnormally short arms and legs.
(Peter Volpe)

complex higher organisms, including humans. Interestingly, the maxim derives overwhelming support from the varied frequencies of the gene for sickle-cell anemia in human populations. The obviously harmful sickling gene may actually confer an advantage to its carriers in certain geographical localities (see chapter 8).

The process of mutation furnishes the genetic variants that are the raw materials of evolution. Ideally, mutations should arise only when advantageous, and only when needed. This, of course, is fanciful thinking. Mutations occur irrespective of their usefulness or uselessness. The mutations responsible for achondroplasia and retinoblastoma (malignant eye tumors) in humans are certainly not beneficial. But novel heritable characters arise repeatedly as a consequence of mutation. Only one mutation in several thousands might be advantageous, but this one mutation might be important, if not necessary, to the continued success of a population. The harsh price of evolutionary potentialities for a population is the continual occurrence and elimination of mutant genes with detrimental effects. Thus, in evolutionary terms, a population, if it is to continue to evolve, must depend on the occasional errors that occur in the copying process of its genetic material.

will be more advantageous than an already established favorable gene is slim. Nonetheless, if the environment were to change, the mutant gene might prove to be beneficial in the new environmental situation. The microscopic water flea, *Daphnia,* thrives at a temperature of 20°C and cannot survive when the temperature rises to 27°C. A mutant strain of this water flea is known that requires temperatures between 25°C and 30°C and cannot live at 20°C. Thus, at high temperatures, the mutant gene is essential to the survival of the water fleas. This little episode reveals an important point: *A mutation that is inferior in the environment in which it arose may be superior in another environment.* Skeptics might contend that this proverbial declaration has no relevancy to

CHROMOSOMAL ABERRATIONS AND PREGNANCY LOSS

Our discussion thus far has concentrated on changes at the molecular level within the gene, the so-called *point* (or *gene*) mutations. Grosser alterations involving large parts of the chromosomes have been termed *chromosomal aberrations.* Some of the chromosomal aberrations are numerical, affecting the amount of chromosomal material. A cell, for example, may have three sets of chromosomes (the triploid state) rather than the customary two chromosomal sets (the standard diploid state). Other gross chromosomal deviations are classified as structural, as represented by the transfer of chromosomal segments between chromosomes (translocation) or the loss of a part of a chromosome (deletion). Once considered exceptionally rare and relatively unimportant, chromosomal aberrations now loom as a prominent cause of the prenatal loss of human embryos.

It may be disconcerting to learn that there is a large natural (spontaneous) loss of human embryos in early pregnancy, a level that far exceeds the incidence in most mammalian species. In the 1950s, A.T. Hertig, a pathologist at the Harvard Medical School, drew attention to the fact that most conceptual losses occur before pregnancy has been diagnosed in the woman. Hertig concluded that only about 40 percent of the early embryos are of such viability as to cause the woman to miss her expected menstrual period.

The data are as yet too imprecise to provide a true rate of pregnancy loss in the human female. The available information, however, does permit a mathematical estimation that appears to reliably approximate natural events. A table of intrauterine death prepared in 1977 by H. Leridon (table 4.3) explores what happens to 100 eggs produced by women who are reproducing naturally. His initial premise is that 16 of the 100 eggs will fail to be fertilized, even under optimal conditions. Of the 84 eggs that are fertilized, 15 will fail to implant. Of the 69 embryos implanted at the end of the first week, 27 of these characteristically will find the uterine lining inhospitable. Thus, up to the second week after ovulation, only 42 eggs of the original 100 will still be viable. Stated another way, by the time pregnancy is recognizable, more than half of the eggs have been lost. At 8 weeks' gestation, when the embryo is termed a fetus, about 65 eggs of the original 100 will have failed to survive. The incidence of spontaneous abortion during the fetal period is very low. At the end of gestation, the probability of a live birth is only 0.31, or 31 percent. For ease of comprehension, the numerical information in table 4.3 is pictorially represented in figure 4.10, in which the baseline is an original 20 eggs rather than 100 eggs.

TABLE 4.3 | **Table of Intrauterine Deaths Per 100 Ova Exposed to the Risk of Fertilization**

Weeks after ovulation	Survivors[a]	Failures[b]
	100	16[c]
0	84[d]	15
1	69[e]	27
2[f]	42	5.0
6	37	2.9
10	34.1	1.7
14	32.4	0.5
18	31.9	0.3
22	31.6	0.1
26	31.5	0.1
30	31.4	0.1
34	31.3	0.1
38	31.2	0.2
Live births	31	

[a] Pregnancies still in progress.
[b] Spontaneous abortions.
[c] Not fertilized.
[d] Number fertilized.
[e] Number implanted.
[f] Expected times of menses.

With the rapid advances in chromosome methodology in the 1960s, the human chromosome complement could be examined in the cells of spontaneous abortuses. To most scientists, the results of the analysis were astonishing. Several thorough investigations have revealed that 50 to 70 percent of first-trimester spontaneous abortuses are chromosomally abnormal. The data demonstrate convincingly that chromosomal aberrations are the major etiologic agents responsible for naturally occurring abortions. We may cite the extensive findings in 1976 by the French team of investigators, J.G. Boué and A. Boué, on the incidence of chromosomal anomalies in human abortuses. Among 1,097 specimens between the second and seventh weeks of gestation, 724, or 66 percent, had an abnormal karyotype (fig. 4.11). The incidence of chromosomal abnormalities fell to 23 percent among the 108 abortuses between 8 and 12 weeks of age.

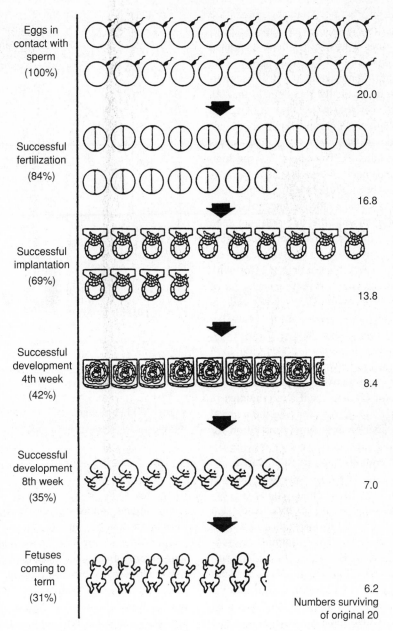

Eggs in contact with sperm (100%) — 20.0

Successful fertilization (84%) — 16.8

Successful implantation (69%) — 13.8

Successful development 4th week (42%) — 8.4

Successful development 8th week (35%) — 7.0

Fetuses coming to term (31%) — 6.2

Numbers surviving of original 20

Figure 4.10 **The fate of 20 eggs** produced by women who are reproducing naturally. Under conditions optimal for fertilization and development, only 6.2 eggs of the original 20 (31%) develop successfully to term.

Most types of chromosomal aberrations are incompatible with survival to an advanced stage of pregnancy. In other words, it has been estimated conservatively that 1 in 5 of all conceptuses carries a chromosome abnormality. At birth, the incidence declines dramatically to about 1 in 200. Based on studies of the mouse, rat, rabbit, and hamster, the frequency of chromosomal abnormalities at conception in these laboratory mammals is far less than in humans. The reason for the high levels of chromosomal aberrations in humans is unknown. It has been suggested that a high frequency of chromosomal abnormalities may confer benefits by lengthening the interval between births. During her lifetime, the human female generally has only a few offspring. It is in the interest of the parents to expend a great deal of effort with each child to ensure the survival and success of each child. The investment of energy per child is likely to be less in situations in which successive births are very close to one another. Conversely, each child is likely to receive greater attention when the time period between births is extended. Accordingly, the provocative hypothesis is that the high frequency of chromosomal abnormalities in humans is nature's way of prolonging the interval between births, thereby reducing competition for parental care and presumably enhancing the fitness of each individual child born and reared.

Is there an "optimal" interval in humans between the delivery of one child and the conception of the next—that is, the interval associated with the greatest probability of giving birth to a normal infant born at full term? Illuminating studies reported in 1999 by Bao-Ping Zhu and his colleagues, in the prestigious *The New England Journal of Medicine*, revealed that a short interval between pregnancies is associated with increased risk of a variety of adverse birth outcomes. The risk of low-birth weight, preterm delivery, or small size

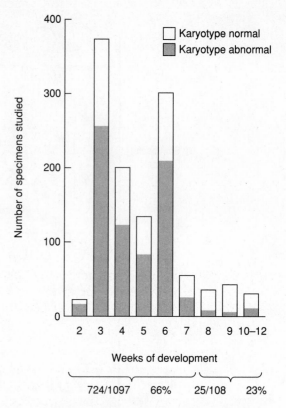

Figure 4.11 Frequency of chromosomal abnormalities in spontaneous abortuses in relation to weeks of development.

for gestational age was 30 to 40 percent higher among infants conceived less than 6 months after a birth than among infants conceived 18 to 23 months after a birth. For infants conceived after a long interpregnancy interval of 120 months, the risk of unfavorable birth outcomes was nearly doubled. The team of investigators concluded that the optimal birth-to-conception interval for safeguarding against adverse birth outcomes was approximately 2 years (specifically, 18 to 23 months).

GENETIC EQUILIBRIUM

5

The opening chapter introduced us to natural populations of bullfrogs that conspicuously contained at one time several hundred multilegged variants. One of the hypotheses offered was that the multilegged anomaly is an inherited condition, transmitted by a detrimental recessive gene. The multilegged bullfrogs disappeared in nature as dramatically as they appeared. They unquestionably failed to reproduce and leave descendants. Now, let us imagine that the multilegged frogs were as reproductively fit as their normal kin. Would the multilegged trait eventually still be eliminated from the population?

A comparable question was posed to the English geneticist R. C. Punnett at the turn of the century. He was asked to explain the prevalence of blue eyes in humans in view of the acknowledged fact that the blue-eyed condition was a recessive characteristic. Would it not be the case that the dominant brown-eyed trait would in time supplant the blue-eyed state in the human population? The answer was not self-evident, and Punnett sought out his colleague Godfrey H. Hardy, the astute mathematician at Cambridge University. Hardy

had only a passing interest in genetics, but the problem intrigued him as a mathematical one. The solution, which we shall consider below, ranks as one of the most fundamental theorems of genetics and evolution. As fate would have it, Hardy's formula was arrived at independently in the same year (1908) by a physician, Wihelm Weinberg, and the well-known equation presently bears both their names.

MENDELIAN INHERITANCE

We may recall that the genetic constitutions, or genotypes, of the normal and multilegged bullfrogs have been designated as AA (normal), Aa (normal but a carrier), and aa (multilegged). The kinds and proportions of offspring that can arise from matings involving the three genotypes are illustrated in figure 5.1. Six different types of matings are possible. The mating $AA \times AA$ gives rise solely to normal homozygous offspring, AA. Two kinds of progeny, AA and Aa in equal proportions, result from the cross of a homozygous normal parent (AA) and a

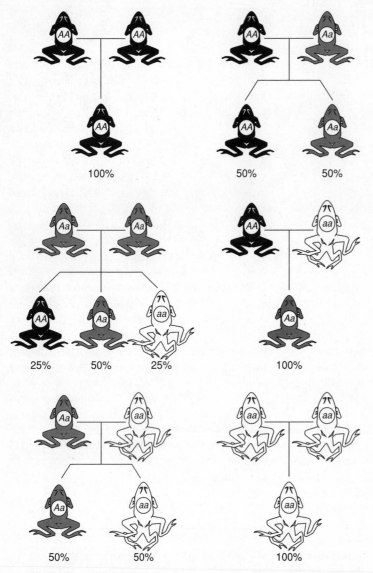

the mating *AA* × *aa*. Both heterozygous *(Aa)* and recessive *(aa)* progeny, in equal numbers, arise from the cross of a heterozygous parent *(Aa)* and a recessive parent *(aa)*. Lastly, two recessive parents *(aa* × *aa)* produce only recessive offspring *(aa)*.

These principles of Mendelian inheritance merely inform us that certain kinds of offspring can be expected from certain types of matings. If we are interested in following the course of a population from one generation to the next, then additional factors enter the scene.

RANDOM MATING

One important factor that influences the genetic composition of a population is the system of mating among individuals. The simplest scheme of breeding activity in a population is referred to as *random mating,* wherein any one individual has an equal chance of pairing with any other individual. Random mating does not mean promiscuity; it simply means that those who choose each other as mating partners do not do so on the basis of similarity or dissimilarity in a given trait or gene.

The absence of preferential mating in a population has interesting consequences. Let

Figure 5.1 Six possible mating types with respect to one pair of genes, and the kinds and percentages of offspring from each type of mating. Normal frogs are either homozygous dominant *(AA)* or heterozygous *(Aa)*; multilegged frogs are recessive *(aa)*. The sex of the parent is not denoted; in crosses of unlike genotypes *(*such as *AA* × *Aa),* either genotype may be the male or the female.

heterozygous parent *(Aa)*. The mating of two heterozygotes *(Aa* × *Aa)* produces *AA, Aa,* and *aa* offspring in the classical Mendelian ratio of 1:2:1. Only heterozygous offspring *(Aa)* emerge from

us suppose that random mating prevails in our particular population of bullfrogs. This assemblage of frogs is ordinarily very large, numbering several thousand individuals. For ease of presentation, however, the size of the population is reduced to 48 males and 48 females. Moreover, for each sex, we may simplify the mathematical computations by assuming that 36 are normal (12 *AA* and 24 *Aa*) and 12 are multilegged *(aa)*. Accordingly, one-quarter of the individuals of each sex are homozygous dominant, one-half are heterozygous, and one-quarter are recessive. Now, if mating occurs at random, will the incidence of multilegged frogs decrease, increase, or remain the same in the next generation?

The problem may be approached by determining how often a given type of mating occurs. Here we will bring into play the multiplication rule of probability: *The chance that two independent events will occur together is the product of their chances of occurring separately.* The proportion of *AA* males in our arbitrary bullfrog population is 1/4. We may also say the chance that a male bullfrog is *AA* is 1/4. Likewise the probability that a female bullfrog is *AA* is 1/4. Consequently, the chances that an *AA* male will "occur together," or mate, with an *AA* female are 1/16 (1/4 × 1/4). The computations for all types of matings can be facilitated by coupling the males and females in a multiplication table, as shown in table 5.1.

Table 5.1 shows that there are nine combinations of mated pairs and that some types occur more frequently than others. It may be helpful to express the frequencies in terms of actual numbers. Thus, for a total of 48 matings, 3 (=1/16 × 48) would be *AA*♀ × *AA*♂, 6 (=1/16 X 48) would

be *AA*♀ × *Aa*♂, 12 (=4/16 × 48) would be *Aa*♀ × *Aa*♂, and so forth. The numbers of each type of mating are listed in table 5.2.

Our next step is to ascertain the kinds and proportions of offspring from each mating. We shall assume that each mated pair yields the same number of offspring—for simplicity, four offspring. (This is an inordinately small number, as a single female bullfrog can deposit well over 10,000 eggs.) We also take for granted that the genotypes of the four progeny from each mating are those that are theoretically possible in Mendelian inheritance (see fig. 5.1). For example, if the parents are *Aa* × *Aa*, their offspring will be 1 *AA*, 2 *Aa*, and 1 *aa*. In another instance, if the parents are *AA* × *Aa*, then the offspring will be 2 *AA* and 2 *Aa*. The outcome of all crosses is shown in table 5.2. It is important to note that the actual numbers of offspring recorded in table 5.2 are related to the frequencies of the different types of matings. For example, the mating of an *AA* female with an *Aa* male occurs six times; hence, the numbers of offspring are increased sixfold

TABLE 5.2	First Generation of Offspring			
Type of Mating (Female × Male)	Number of Each Type of Mating*	Number of Offspring		
		AA	*Aa*	*aa*
AA × *AA*	3	12		
AA × *Aa*	6	12	12	
AA × *aa*	3		12	
Aa × *AA*	6	12	12	
Aa × *Aa*	12	12	24	12
Aa × *aa*	6		12	12
aa × *AA*	3		12	
aa × *Aa*	6		12	12
aa × *aa*	3	—	—	12
		48	96	48
		(25%)	(50%)	(25%)

* Based on a total of 48 matings.

TABLE 5.1	Random Mating of Individuals		
	Male (♂)		
Female (♀)	1/4 *AA*	2/4 *Aa*	1/4 *aa*
1/4 *AA*	1/16 *AA* × *AA*	2/16 *AA* × *Aa*	1/16 *AA* × *aa*
2/4 *Aa*	2/16 *Aa* × *AA*	4/16 *Aa* × *Aa*	2/16 *Aa* × *aa*
1/4 *aa*	1/16 *aa* × *AA*	2/16 *aa* × *Aa*	1/16 *aa* × *aa*

(from 2 each of *AA* and *Aa* to 12 each of the two genotypes).

An examination of table 5.2 reveals that the kinds and proportions of individuals in the new generation of offspring are exactly the same as in the parental generation. There has been no change in the ratio of normal frogs (75 percent *AA* and *Aa*) to multilegged frogs (25 percent *aa*). In fact, the proportions of phenotypes (and genotypes) will remain the same in all successive generations, provided that the system of random mating is continued.

GENE FREQUENCIES

There is a less tedious method of arriving at the same conclusion. Rather than figure out all the matings that can possibly occur, we need only to consider the genes (alleles) that are transmitted by the eggs and sperm of the parents. Let us assume that each parent produces only 10 gametes. The 12 homozygous dominant males *(AA)* of our arbitrary initial population can contribute 120 sperm cells to the next generation, each sperm containing one *A* allele. The 24 heterozygous males *(Aa)* can transmit 240 gametes, 120 of them with *A* and 120 with *a*. The remaining 12 recessive males *(aa)* can furnish 120 gametes, each with an *a* allele. The total pool of alleles provided by all males will be 240 *A* and 240 *a*, or 50 percent of each kind. Expressed as a decimal fraction, the frequency of allele *A* is 0.5; of *a*, 0.5.

Since the females in our population have the same genetic constitutions as the males, their gametic contribution to the next generation will also be 0.5 *A* and 0.5 *a*. The eggs and sperm can now be united at random in a genetical checkerboard (fig 5.2).

It should be evident from figure 5.2 that the distribution of genotypes among the offspring is 0.25 *AA* : 0.50 *Aa* : 0.25 *aa*. The random union of eggs and sperm yields the same result as the random mating of parents (refer to table 5.2). Thus, using two different approaches, we have answered the question posed in the introductory remarks to this chapter. *If the multilegged frogs are equally as fertile as the normal frogs and leave equal numbers of offspring each generation, then these anomalous frogs will persist in the population with the same frequency from one generation to the next.*

HARDY-WEINBERG EQUILIBRIUM

A population in which the proportions of genotypes remain unchanged from generation to generation is in *equilibrium*. The fact that a system of random mating leads to a condition of equilibrium was uncovered independently by G. H. Hardy and W. Weinberg, and has come to be widely known as the *Hardy-Weinberg equilibrium*. This theorem states that the proportions of *AA, Aa,* and *aa* genotypes, as well as the proportions of *A* and *a* alleles, will remain constant from generation to generation, provided that the bearers of the three genotypes have equal opportunities of producing offspring in a large, randomly mating population.

The above statement can be translated into a simple mathematical expression. If we let *p* be the frequency of allele *A* in the population, and *q*

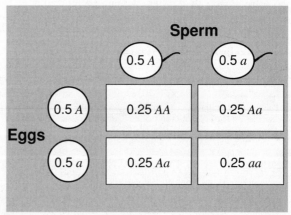

Figure 5.2 Random union of eggs and sperm yeilds the same outcome as the random mating of parents (refer to table 5.2).

equal the frequency of the alternative allele, *a*, then the distribution of the genotypes in the next generation will be $p^2AA : 2pq\ Aa : q^2\ aa$. This relationship may be verified by the use, once again, of a genetical checkerboard (fig. 5.3). Mathematically inclined readers will recognize that $p^2 : 2pq : q^2$ is the algebraic expansion of the binomial $(p + q)^2$. The frequencies of the three genotypes (0.25 *AA* : 0.50 *Aa* : 0.25 *aa*) in our bullfrog population under the system of random mating is the expanded binomial $(0.5 + 0.5)^2$.

We may consider another arbitrary population in the equilibrium state. Suppose that the population consists of 16 *AA,* 48 *Aa,* and 36 *aa* individuals. We may assume, as before, that 10 gametes are contributed by each individual to the next generation. All the gametes transmitted by the 16 *AA* parents (numerically, 160) will contain the *A* allele, and half the gametes (240) provided by the *Aa* parents will bear the *A* allele. Thus, of the 1,000 total gametes in the population, 400 will carry the *A* allele. Accordingly, the frequency of allele *A* is 0.4 (designated *p*). In like manner, it can be shown that the frequency of allele a is 0.6 (*q*). Substituting the numerical values for *p* and *q* in the Hardy-Weinberg formula, we have:

$p^2\ AA : 2pq\ Aa : q^2aa$

$(0.4)^2\ AA : 2(0.4)\ (0.6)\ Aa : (0.6)^2\ aa$

0.16 *AA* : 0.48 *Aa* : 0.36 *aa.*

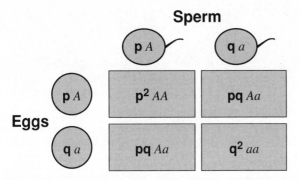

Sperm

Figure 5.3 **The distribution of genotypes** in the next generation is $p^2\ AA : 2pq : Aa : q^2aa$ (Hardy-Weinberg formula).

Hence, the proportions of the three genotypes are the same as those of the preceding generation.

It should also be clear that a recessive trait, such as blue eyes in humans, will not become rare just because it is governed by a recessive gene. Nor can the dominant brown-eyed condition become widespread simply by virtue of its dominance. Whether a given gene is common or rare is controlled by other factors, particularly natural selection.

ESTIMATING THE FREQUENCY OF HETEROZYGOTES

It comes as a surprise to many readers to discover that the heterozygotes of a rare recessive abnormality are rather common instead of being comparatively rare. Recessive albinism may be used as an illustration. The frequency of albinos is about 1/20,000 in human populations. When the frequency of the homozygous recessive (q^2) is known, the frequency of the recessive allele (*q*) can be calculated, as follows:

$q^2 = 1/20,000 = 0.00005$
$q = \sqrt{0.00005} = 0.007$
$= \text{about } 1/140$
(frequency of recessive allele).

The heterozygotes are represented by *2pq* in the Hardy-Weinberg formula. Accordingly, the frequency of heterozygous carriers of albinism can be calculated as follows:

$q = 0.007$
$p = 1 - 0.007 = 0.993.$
$\therefore 2pq = 2(0.993 \times 0.007) = 0.014$
$= \text{about } 1/70 \text{ (frequency of}$
heterozygote).

Thus, although 1 person in 20,000 is an albino (recessive homozygote), about 1 person in 70 is a heterozygous carrier. There are 280 times as many carriers as affected individuals! It bears emphasizing that the rarity of a recessive disorder does not signify a comparable rarity of heterozygous carriers. In fact, when the frequency of the recessive gene is extremely low, nearly all the recessive genes are in the heterozygous state.

IMPLICATIONS

The Hardy-Weinberg theorem is *entirely hypothetical*. The set of underlying assumptions can scarcely be fulfilled in any natural population. We implicitly assume the absence of recurring mutations, the absence of any degree of preferential matings, the absence of differential mortality or fertility, the absence of immigration or emigration of individuals, and the absence of fluctuations in gene frequencies due to sheer chance. But therein lies the significance of the Hardy-Weinberg theorem. In revealing the conditions under which evolutionary change cannot occur, it brings to light the possible forces that could cause a change in the genetic composition of a population. The Hardy-Weinberg theorem thus depicts a static situation.

An understanding of Hardy-Weinberg equilibrium provides a basis for recognizing the forces that permit evolutionary change. A few of the more obvious factors that prevent a natural population from attaining stationary equilibrium are: (1) mutation, (2) natural selection, (3) chance events in a small population (genetic drift), and (4) migration. These factors or forces may profoundly modify the gene frequencies in natural populations. In essence, the Hardy-Weinberg theorem represents the cornerstone of population genetic studies, since deviations from the Hardy-Weinberg expectations direct attention to the evolutionary forces that upset the theoretical expectations. Evolution can be plausibly defined as the disturbance of, or shift in, the Hardy-Weinberg equilibrium.

CONCEPT OF SELECTION

W e have remarked that the multilegged anomaly which arose in a local bullfrog population in Mississippi in 1958 has not been detected since its initial occurrence. One of our suppositions was that the multilegged trait is governed by a recessive mutant gene. We might presume that the mutant gene responsible for the abnormality has disappeared entirely from the population. But can a detrimental mutant gene be completely eradicated from natural populations of organisms, even in the face of the severest form of selection? Most persons are frankly puzzled when they are informed that the answer is no. Yet, our knowledge of the properties of mutation and selection expressly permits a firm negative reply.

SELECTION AGAINST RECESSIVE DEFECTS

In the preceding chapter, in our consideration of the Hardy-Weinberg equilibrium, we assumed that the mutant multilegged frogs *(aa)* were as reproductively fit as their normal kin *(AA* and *Aa)* and

left equal numbers of living offspring each generation. Now, however, let us presume that all multilegged individuals fail to reach sexual maturity generation after generation. Will the incidence of the multilegged trait decline to a vanishing point?

We may start with the same distribution of individuals in the initial generation as postulated in chapter 5, namely, 24 *AA,* 48 *Aa,* and 24 *aa,* with the sexes equally represented. Since the multilegged frogs *(aa)* are unable to participate in breeding, the parents of the next generation comprise only the 24 *AA* and 48 *Aa* individuals. The heterozygous types are twice as numerous as the homozygous dominants; accordingly, two-thirds of the total breeding members of the population are *Aa* and one-third are *AA.* We may once again employ a genetic checkerboard (table 6.1) to ascertain the different types of matings and their relative frequencies.

The frequencies of the different matings shown in table 6.1 may be expressed as whole numbers rather than fractions. Given a total of 36 matings, 4 (= 1/9 × 36) would be *AA* ♀ × *AA* ♂, 8 (= 2/9 × 36) would be *AA* ♀ × *Aa* ♂, and 16

(= 4/9 × 36) would be *Aa* ♀ × *Aa* ♂. These numbers are recorded in table 6.2.

Our next task is to determine the outcome of each type of cross. We shall assume that each mated pair contributes an equal number of progeny to the next generation (say, four offspring). As revealed in table 6.2, the offspring are distributed according to Mendelian ratios, and the actual numbers of offspring reflect the frequencies of the different kinds of matings. For example, a single *AA* ♀ × *Aa* ♂ mating yields 4 offspring in the Mendelian ratio of 2 *AA* : 2 *Aa*. There are, however, eight matings of this kind; the numbers of offspring are correspondingly increased to 16 *AA* and 16 *Aa*.

Even though all the multilegged frogs fail to reproduce, the detrimental recessive genes are still transmitted to the first generation. The emergence of multilegged frogs in the first generation stems from the matings of two heterozygous frogs. However, as seen from table 6.2, the frequency of the multilegged trait *(aa)* decreases from 25 percent to 11.11 percent in a single generation.

The effects of complete selection against the multilegged frogs in subsequent generations can be determined by the foregoing method of calculation, but the lengthy tabulations can be wearisome. At this point we may apply a formula that will establish in a few steps the frequency of the recessive allele after any number of generations of complete selection:

$$q_n = \frac{q_o}{1 + nq_o}.$$

In the above expression, q_o represents the initial or original frequency of the recessive allele, and q_n is the frequency after n generations. Thus, with the initial value $q_o = 0.5$, the frequency of the recessive allele after two generations ($n = 2$) will be:

$$q_2 = \frac{q_o}{1 + 2q_o} = \frac{0.5}{1 + 2(0.5)} = \frac{0.5}{2.0} = 0.25.$$

If the frequency of the recessive allele *(a)* is q, then the frequency of the recessive individual *(aa)* is q^2. Accordingly, the frequency of the recessive homozygote is $(0.25)^2$, or 0.0625 (6.25 percent). In the second generation, therefore, the incidence of the multilegged trait drops to 6.25 percent.

If we perform comparable calculations through several generations, we emerge with a comprehensive picture that is tabulated in table 6.3 and portrayed in figure 6.1. In the third generation, the frequency of the recessive homozygote declines to 4.0 percent. Progress in terms of the elimination of the multilegged trait is initially rapid but becomes slower as selection is continued over many successive generations. About 20 generations are required to depress the incidence of the multilegged trait to 2 in 1,000 individuals

TABLE 6.1	Matings and Relative Frequencies

Female (♀)	Male (♂)	
	1/3 AA	2/3 Aa
1/3 AA	1/9 AA × AA	2/9 AA × Aa
2/3 Aa	2/9 Aa × AA	4/9 Aa × Aa

TABLE 6.2	First Generation of Offspring After Complete Selection

Type of Mating (Female × Male)	Number of Each Type of Mating*	Number of Offspring		
		AA	Aa	aa
AA × AA	4	16		
AA × Aa	8	16	16	
Aa × AA	8	16	16	
Aa × Aa	16	16	32	16
		64	64	16
		(44.44%)	(44.44%)	(11.11%)

* Based on a total of 36 matings.

TABLE 6.3	Effects of Complete Selection Against a Recessive Trait

Generations	Recessive Allele Frequency	Recessive Homozygotes %	Heterozygotes %	Dominant Homozygotes %
0	0.500	25.00	50.00	25.00
1	0.333	11.11	44.44	44.44
2	0.250	6.25	37.50	56.25
3	0.200	4.00	32.00	64.00
4	0.167	2.78	27.78	69.44
8	0.100	1.00	18.00	81.00
10	0.083	0.69	15.28	84.03
20	0.045	0.20	8.68	91.12
30	0.031	0.10	6.05	93.85
40	0.024	0.06	4.64	95.30
50	0.020	0.04	3.77	96.19
100	0.010	0.01	1.94	98.05

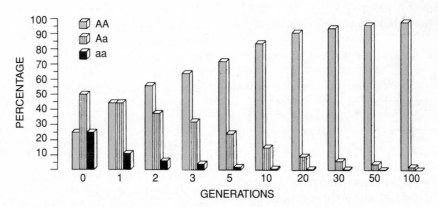

Figure 6.1 Effects of complete selection against recessive homozygotes *(aa)* occurring initially ("0" generation) at a frequency of 25 percent. The effectiveness of selection in reducing the incidence of the recessive trait decreases with successive generations. The frequency of recessive homozygotes drops markedly from 25 percent to 6.25 percent in two generations. However, 8 generations are required to reduce the incidence of the recessive trait to 1.0 percent, 30 generations are needed to achieve a reduction to 0.1 percent, and approximately 100 generations to depress the frequency to 0.01 percent.

(0.20 percent). Ten additional generations are necessary to effect a reduction to 1 in 1,000 individuals (0.10 percent). Thus, as a recessive trait becomes rarer, selection against it becomes less effective. The reason is quite simple: very few recessive homozygotes are exposed to the action of selection. The now rare recessive allele *(a)* is carried mainly by heterozygous individuals *(Aa),* where it is sheltered from selection by its normal dominant partner *(A).*

SIGNIFICANCE OF THE HETEROZYGOTE

When the frequency of a detrimental recessive gene becomes very low, most affected offspring *(aa)* will come from matings of two heterozygous carriers *(Aa)*. For example, in the human population, the vast majority of newly arising albino individuals *(aa)* in a given generation (more than 99 percent of them) will come from normally pigmented heterozygous parents. Considerations of this kind led us to postulate in chapter 1 that the multilegged frogs that appeared suddenly in the natural population were derived from normal-legged heterozygous parents (refer to fig. 1.5).

Detrimental recessive genes in a population are unquestionably harbored mostly in the heterozygous state. As shown in table 6.4, the frequency of heterozygous carriers is many times greater than the frequency of homozygous individuals afflicted with a trait. Thus, an extremely rare disorder like alkaptonuria (blackening of urine) occurs in 1 in one million persons. This detrimental gene, however, is carried in the hidden state by 1 out of 500 persons. There are 2,000 times as many genetic carriers of alkaptonuria as there are individuals affected with this defect. For another recessive trait, cystic fibrosis, 1 out of 2,500 individuals is affected with this homozygous trait. One of 25 persons is a carrier of cystic fibrosis. In modern genetic counseling programs, an important consideration has been the development of simple, inexpensive means of detecting heterozygous carriers of inherited disorders.

INTERPLAY OF MUTATION AND SELECTION

Ideally, if the process of complete selection against the recessive homozygote were to continue for several hundred more generations, the detrimental recessive gene would be completely eliminated and the population would consist uniformly of normal homozygotes *(AA)*. But, *in reality,* the steadily diminishing supply of deleterious recessive genes is

TABLE 6.4	Frequencies of Recessive Homozygotes and Heterozygous Carriers		
Frequency of Homozygotes *(aa)*	**Frequency of Heterozygous Carriers** *(Aa)*	**Ratio of Carriers to Homozygotes**
1 in 500 (Sickle-Cell Anemia)[a]	1 in 10	50:1
1 in 2,500 (Cystic Fibrosis)	1 in 25	100:1
1 in 6,000 (Tay-Sachs Disease)[b]	1 in 40	150:1
1 in 20,000 (Albinism)	1 in 70	286:1
1 in 25,000 (Phenylketonuria)	1 in 80	313:1
1 in 50,000 (Acatalasia)[c]	1 in 110	455:1
1 in 1,000,000 (Alkaptonuria)	1 in 500	2,000:1

[a]Based on incidence among African Americans. Sickle-cell anemia is later described (chapter 8) as a co-dominant trait.

[b]In the United States, the disease occurs once in 6,000 Jewish births and once in 500,000 non-Jewish births.

[c]Based on prevalence rate among the Japanese.

continually being replenished by recurrent mutations from normal *(A)* to abnormal *(a)*. Mutations from *A* to *a,* which inevitably occur from time to time, were not taken into account in our determinations. Mutations, of course, cannot be ignored.

All genes undergo mutations at some definable rate. If a certain proportion of *A* alleles are converted into *a* alleles in each generation, the population will at all times carry a certain amount of the recessive mutant gene *(a)* despite selection against it. Without any sophisticated calculations, it can be shown that a point will be reached at which the number of the variant recessive alleles eliminated by selection just balances the number of the same variant recessive alleles produced by mutation. An analogy shown in figure 6.2A will help in visualizing this circumstance. The water level in the beaker remains constant when the rate at which water enters the opening of the beaker equals the rate at which it leaves the hole in the side of the beaker. In other words, a state of equilibrium is reached when the rate at which the recessive gene is replenished by mutation equals the rate at which it is lost by selection. It should be clear that it is not mutation alone that governs the incidence of deleterious recessives in a population. The generally low frequency of harmful recessive genes stems from the dual action of mutation and selection. The mutation process tends to increase the number of detrimental recessives; the selection mechanism is the counteracting agent.

What would be the consequences of an increase in the mutation rate? Humans today live in an environment in which high-energy radiation promotes a higher incidence of mutations. We may return to our analogy (fig 6.2*B*). The increased rate of mutation may be envisioned as an increased input of water. The water level in the beaker will rise and water will escape more rapidly through the hole in the side of the beaker. Similarly, mutant genes will be found more frequently in a population, and they will be eliminated at a faster rate from the population. As before, a balance will be restored eventually between mutation and selection, but now the population has a larger store of deleterious genes and a larger number of afflicted individuals arising each generation.

The supply of defective genes in the human population has already increased through the greater medical control of recessive disorders. The outstanding advances in modern medicine have served to prolong the lives of genetically defective individuals who might otherwise not have survived

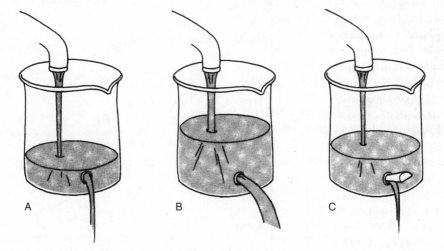

Figure 6.2 **Interplay of detrimental mutant genes** (water from faucet) and their elimination by selection (water escaping through hole) in a population (beaker) containing a pool of the harmful genes (water in beaker). *A.* State of genetic equilibrium (constant water level in beaker) when the rates at which water enters and leaves the beaker are equal. *B.* Effect on an increase in mutation rate (increased flow of faucet water) as might be expected from the continued widespread use of ionizing radiation. A new equilibrium (new constant water level) is established, but the frequency of the detrimental gene in the population is higher (higher water level in beaker). *C.* Effect of reducing selection pressure (decreased exit of water) as a consequence of improving the reproductive fitness of genetically defective individuals by modern medical practices. Mutation rate (inflow of water) is the same as in *A.* The inevitable result is a greater incidence of the harmful mutant gene in the population (higher level of water in the beaker).

to reproductive age. This may be compared to partially plugging the hole in the side of the beaker (fig. 6.2C). The water level in the beaker will obviously rise, as will the amount of deleterious genes in a population. Evidently, the price of our humanitarian principles is the enlargement of our pool of detrimental genes.

PARTIAL SELECTION

We have treated above the severest form of selection against recessive individuals. Complete, or 100 percent, selection against a recessive homozygote is often termed *lethal* selection, and the mutant gene is designated as a *lethal* gene. A lethal gene does not necessarily result in the death of the individual but does effectively prevent the individual from reproducing or leaving offspring. Not all mutant genes are lethal; in fact, the majority of them have less drastic effects on viability or fertility. A mildly handicapped recessive homozygote may reproduce but may be inferior in fertility to the normal individual. When the reproductive capacity of the recessive homozygote is only half as great as the normal type, he or she is said to be *semi-sterile,* and the mutant gene is classified as *semi-lethal.* A *subvital* recessive gene is one that, in double dose, impairs an individual to the extent that his or her reproductive fitness is less than 100 percent but more than 50 percent of normal proficiency.

The action of selection varies correspondingly with the degree of detrimental effect of the recessive gene. Figure 6.3 shows the results of different intensities of selection in a population that initially contains 1.0 percent recessive homozygotes. With complete (lethal) selection, a reduction in the incidence of the recessive trait from 1.0 percent to 0.25 percent is accomplished in 10 generations. Twenty generations of complete selection reduces the incidence to 0.11 percent. When the recessive gene is semi-lethal (50 percent selection), 20 generations, or twice as many generations as under complete selection, are required to depress the frequency of the recessive homozygote to about 0.25 percent. Selection against a subvital gene (for example, 10 percent selection) results in a considerably slower rate of elimination of the recessive homozygotes. When the homozygote is at a very slight reproductive disadvantage (1.0 percent selection), only

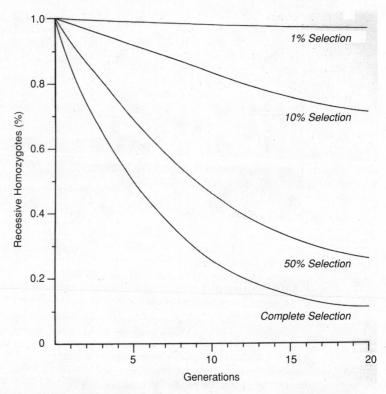

Figure 6.3 Different intensities of selection against recessive homozygotes occurring initially ("0" generation) at a frequency of 1.0 percent. The elimination of recessive individuals per generation proceeds at a slower pace as the strength of selection decreases.

a small decline of 0.03 percent (from 1.0 percent to 0.97 percent) occurs after 20 generations. It is evident that mildly harmful recessive genes may remain in a population for a long time.

These considerations are shown in table 6.5 also. Here, the results are expressed in terms of the value *s,* or the *selection coefficient.* The selection coefficient is a measure of the contribution of one genotype relative to the contributions of the other genotypes. Thus, an *s* value of 1 means that an individual leaves no offspring; the recessive homozygote is lethal. An *s* value of 0.10 signifies that the *aa* homozygote contributes only 90 offspring to the next generation as compared with *AA* and *Aa* individuals, each of whom contributes 100 offspring. It may be noticed, once again, that the rate of decline of the recessive homozygotes becomes slower as the selection coefficient decreases in value.

SELECTION AGAINST DOMINANT DEFECTS

If complete selection acts against an abnormal trait caused by a dominant allele *(A)* so that none of the *AA* or *Aa* individuals leave any progeny, than all the *A* genes are at once eliminated. In the absence of recurrent mutation, all subsequent generations will consist exclusively of homozygous recessive *(aa)* individuals.

TABLE 6.5	Declines in Frequencies of Recessive Homozygotes Based on Different Values of the Selection Coefficient(s)

Generations	s = 1.0	s = 0.50	s = 0.10	s = 0.01
0	1.00*	1.00	1.00	1.00
10	0.25	0.46	0.84	0.98
20	0.11	0.26	0.71	0.97

* The initial frequency of the recessive homozygote in all cases is 1 percent, as in figure 6.3.

However, we must contend again with ever-occurring mutations and the effects of partial selection. The late geneticist Curt Stern provides us with a simple, clear model of this situation. Imagine a population of 500,000 individuals, all of whom are initially homozygous recessive *(aa).* Thus, no detrimental dominant genes *(A)* are present and the population as a whole contains 1,000,000 recessive genes *(a).* In the first generation, 10 dominant mutant genes arise as a result of the recessive gene's mutating to the dominant state at a rate of 1 in 100,000 genes. We shall now assume that the dominant mutant gene is semilethal; in other words, only 5 of the newly arisen dominant genes are transmitted to the next, or second, generation. For ease of discussion, this is pictorially shown in figure 6.4 and represented also in table 6.6. It can be seen that the second generation would contain a total of 15 dominant genes—5 brought forth from the first generation and 10 new ones added by mutation. In the third generation, the 5 dominant genes carried over from the first generation would be reduced to 2.5, the 10 dominant alleles of the second generation would be depressed to 5, and 10 new abnormal alleles would arise anew by mutation. The total number of dominant genes would increase slightly with each subsequent generation, until a point is reached (about 12 generations) where the rate of elimination of the abnormal dominant gene balances the rate of mutation. In other words, the inflow of new dominant alleles by mutation each generation is balanced by the outflow or elimination of the dominant genes each generation by selection.

The equilibrium frequency of a detrimental dominant gene in a population can be altered by changing the rate of loss of the gene in question. In humans, *retinoblastoma,* or cancer of the eye in newborn babies, has until recently been a fatal condition caused by a dominant mutant gene. With modern medical treatment, approximately 70 percent of the afflicted infants can be saved. The effect of increasing the reproductive fitness of the survivors is to raise the frequency of the detrimental dominant gene in the human population. The

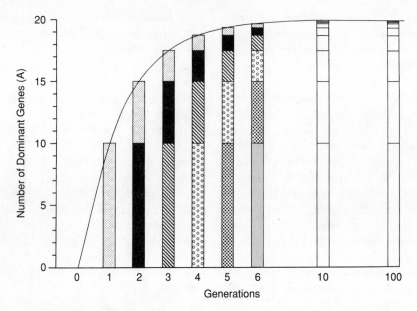

Figure 6.4 Establisment of a constant level of a semilethal dominant gene
(A) in a population over the course of several generations. The fixed number of
new dominant genes introduced each generation through mutations from *a* to *A*
eventually exactly balances the number of dominant genes selectively
eliminated each generation. In this particular case, an equilibrium is reached
(after about 12 generations) when the total number of dominant genes is
approximately 20.

TABLE 6.6	Equilibrium Frequency of a Dominant Gene
	Conditions: 1. Size of Population: 500,000 Individuals
	2. Mutation Rate *(a ⟶ A)*: 1 in 100,000
	3. Selection Coefficient (s) : 0.5 (semilethal)

		Dominant Gene *A*		
Generation	Recessive Gene *a*	Left Over from Former Generations	Newly Mutated	Total
0	1,000,000	—	—	—
1	1,000,000*	—	10	10
2	1,000,000	5	10	15
3	1,000,000	5 + 2.5	10	17.5
4	1,000,000	5 + 2.5 + 1.25	10	18.75
5	1,000,000	5 + 2.5 + 1.25 + 0.625	10	19.375
∞	1,000,000	5 + 2.5 + 1.25 + 0.625 + 0.3125 + ...	10	20.00

*The total number of recessive genes should be reduced by the total number of dominant genes each generation, but this minor correction would be
inconsequential.

accumulation of deleterious genes in the human gene pool has been a matter of awareness and concern.

CONCEALED VARIABILITY IN NATURAL POPULATIONS

From what we have already learned, we should expect to find in natural populations a large number of deleterious recessive genes concealed in the heterozygous state. It may seem that this expectation is based more on theoretical deduction than on actual demonstration. This is not entirely the case. Penetrating studies by a number of investigators of several species of the fruit fly *Drosophila* have unmistakably indicated an enormous store of recessive mutant genes harbored by individuals in nature. We may take as an illustrative example the kinds and incidence of recessive genes detected in *Drosophila pseudobscura* from California populations. The following data are derived from the studies of the late geneticist Theodosius Dobzhansky.

Flies were collected from nature, and a series of elaborate crosses were performed in the laboratory to yield offspring in which one pair of chromosomes carried an identical set of genes. The formerly hidden recessive genes in a given pair of chromosomes were thus all exposed in the homozygous state. All kinds of recessive genes were uncovered in different chromosomes, as exemplified by those unmasked in one particular chromosome, known simply as "the second." About 33 percent of the second chromosomes harbored one or more recessive genes that proved to be lethal or semilethal to flies carrying the second chromosome in duplicate. An astonishing number of second chromosomes—93 percent—contained genes that produced subvital or mildly incapacitating effects when present in the homozygous state. Other unmasked recessive genes resulted in sterility of the flies or severely retarded the developmental rates of the flies. All these flies were normal in appearance when originally taken from nature. It is apparent that very few, if any, outwardly normal flies in natural populations are free of hidden detrimental recessive genes.

GENETIC LOAD IN HUMAN POPULATIONS

The study of the concealed variability, or genetic load, in humans cannot be approached, for obvious reasons, by the experimental breeding techniques used with fruit flies. Estimates of the genetic load in the human population have been based principally on the incidence of defective offspring from marriages of close relatives (consanguineous marriages). It can be safely stated that every human individual contains at least one newly mutated gene. It can also be accepted that any crop of gametes contains, in addition to one or more mutations of recent origin, at least 10 mutant genes that arose in the individuals of preceding generations and that have accumulated in the population. The average person is said to harbor four concealed lethal genes, each of which, if homozygous, is capable of causing death between birth and maturity. The most conservative estimates place the incidence of deformities to detrimental mutant genes in the vicinity of 2 per 1,000 births. *It is evident that humans are not exempt from their share of detrimental genes.*

SELECTION RELAXATION IN HUMANS

Modern medicine, by finding ways to keep alive individuals who carry deleterious genes, encourages the survival of those whom more than likely will pass on their defects to future generations. The conventional ethics of medicine are shaken when, with increasing knowledge, it becomes clear that saving the life of a child with a hereditary disorder also ensures the retention and increase of detrimental genes that natural selection ordinarily keeps at very low frequencies. Can we continue indefinitely to load our population with hereditary disabilities?

Some geneticists have looked with grave concern upon this situation. In particular, the late Nobel laureate Hermann J. Muller was most distressed about the impending genetic disintegration of the human species. Throughout his career, Muller was a persuasive and articulate prophet of doom. Most geneticists today do not share Muller's pessimism and, in fact, consider his gloomy forecast as groundless.

Calculations indicate that the increase in the frequency of detrimental recessive genes due to the relaxation of selection by medical therapy would be extremely slow. For example, given the conditions presented in table 6.7, it would take well over 100 generations (or 3,000 years) for a recessive trait to double in frequency from 1 in 100,000 to 2 in 100,000. If the recessive defect occurred initially with a frequency of 1 in 10,000, then slightly more than 30 generations would be required to double the incidence. In essence, although relaxation of selection tends to engender an increase in the incidence of genetic disorders in future generations, the increase occurs very slowly.

GENERAL EFFECTS OF SELECTION

In our discussion thus far, we have seen that the selection process favors individuals who are best adapted to new situations or to new ecological opportunities. Such selection is said to be *directional* since the norm for the population is shifted with time in one direction (fig. 6.5A). The selection for melanic varieties of moths in industrial areas (see chapter 7) exemplifies directional selection. The curve of distribution shifts to the right (fig. 6.5A) as the darker varieties supplant the formerly abundant light-colored moths.

The selection process may take other forms. *Stabilizing selection* acts to reduce the array of gene complexes that can be expressed in a population. As seen in figure 6.5B, the shape of the curve tends to narrow through the continual elimination of the less-adapted individuals at the extremes of the distribution curve. Stated differently, the intermediate values for a given trait are favored over the extreme values. The birth weight of newborns provides an instructive example of a human characteristic that has been subjected to stabilizing selection (fig. 6.6). The optimum birth weight is 7.3 pounds; newborn infants less than 5.5 pounds and greater than 10 pounds have the highest probability of mortality. Given the strong stabilizing influence of weeding out the extremes, the optimum birth weight is associated with the lowest mortality. The curve for mortality is virtually the complement of the curve for survival.

Disruptive selection is the most unusual of the three types of selection (fig. 6.5C). This form of selection occurs when the extreme values have the highest fitness and the intermediate values are

TABLE 6.7	**Selection Relaxation for a Formerly Lethal Recessive Whose Fitness Is Restored to 100 Percent by Medical Therapy** **Conditions: 1. Initial Frequency of Recessive Defect at Birth: 1/100,000** **(1×10^{-5})** ** 2. Mutation Rate $(A \longrightarrow a)$: 1/100,000 (1×10^{-5})**

Generation	Frequency of Recessive Defect at Birth
0	1×10^{-5}
1	1.01×10^{-5}
3	1.02×10^{-5}
10	1.06×10^{-5}
30	1.20×10^{-5}
100	1.73×10^{-5}

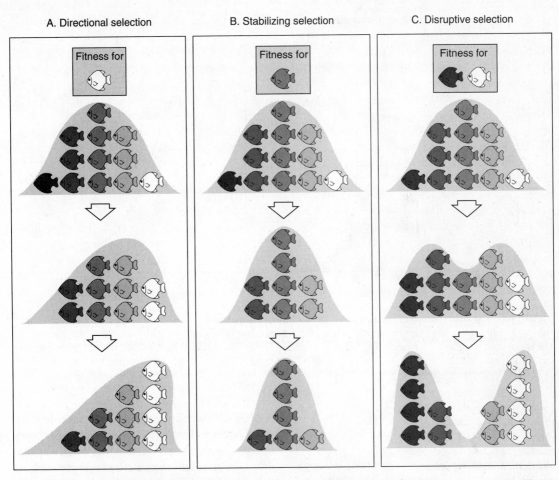

Figure 6.5 Schematic representation of three types of selection and their effects. Each curve represents the normal distribution of the trait in a population. From top to bottom, the lower curves show the expected distribution after the impact of selection. *(A) Directional selection.* The adaptive norm changes as less adapted genotypes are replaced by better adapted genotypes. *(B) Stabilizing selection.* The intermediate values for a given trait are favored (preserved) and the shape of the curve narrows through the elimination of the carriers of the extreme values. *(C) Disruptive selection.* Two adaptive norms are generated when the population exists in a heterogeneous environment.

relatively disadvantageous in terms of reproductive effectiveness. It is, essentially, selection for diversification with respect to a trait. The shell patterns of limpets (marine molluscs) form a continuum ranging from pure white to dark tan. Limpets typically dwell in one of two distinct habitats, attaching to either white gooseneck barnacles or tan-colored rocks. As might be expected, the light-colored limpets seek the protection of the white

barnacles, whereas the tan limpets live by choice almost exclusively on the dark rocks. Limpets of intermediate shell patterns are conspicuous and are intensely selected against by predatory shore birds. If this disruptive type of selection (favoring the extremes) were to be accompanied by the sexual isolation of the two types of limpets, two new species could arise. The mode of origin of species is treated in detail in a subsequent chapter.

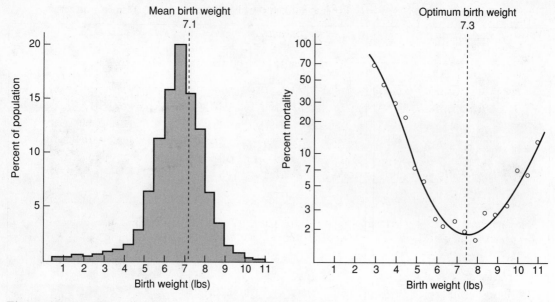

Figure 6.6 The distribution of birth weights of human newborns and the mortality of the various birth-weight classes. The histogram *(left)* shows the proportions of the population falling into the various birth-weight classes. The mean birth weight is 7.1 pounds. The curve of mortality *(right)* in relation to birth weight reveals that the lowest mortality is associatted with the optimum birth weight (7.3 pounds). (Based on data by Karn and Penrose, 1951.)

NATURAL SELECTION AND PREGNANCY LOSS

We remarked earlier (chapter 4) that there is a large natural loss of human embryos in pregnancy. The important consideration now is the high efficiency with which nature eliminates chromosomally abnormal embryos during the course of pregnancy. For every 1,000 chromosomal abnormalities that are present in embryos in the uterus, only 6 are expected to survive to the point of a live birth. Thus 99.4 percent of the chromosomal abnormalities are eliminated naturally through spontaneous abortion.

We may direct our attention to the 0.6 percent of newborn infants affected with a chromosomal abnormality. As seen in figure 6.7, some members of this affected group comprise the XO (Turner) and XXY (Klinefelter) conditions. These individuals are effectively eliminated from the reproductive

pool by their sterility. In contrast, semisterile carriers of balanced translocation tend to be normal phenotypically but are at risk of having abnormal children. The risk, however, is reduced appreciably by the segregation of unbalanced chromosome complements in gametes that are likely to be inviable.

From the perspective of natural selection, the XXX females and the XYY males represent an instructive group. Both types of individuals are usually fertile, and their children are chromosomally normal (XX daughters and XY sons). Some mechanism operates during meiosis to eliminate the extra X chromosome from the egg, or the extra Y chromosome from the sperm. Thus, natural selection has modified the meiotic divisions to ensure that only normal haploid gametes are produced by XXX females and XXY males.

The salient feature of the foregoing considerations is the efficacy with which natural selection eliminates chromosomally abnormal conceptuses—

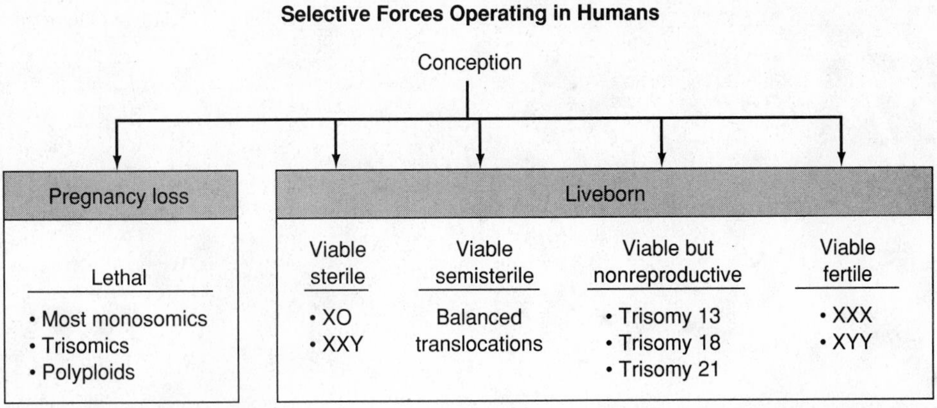

Selective Forces Operating in Humans

Conception

Pregnancy loss	Liveborn			
	Viable sterile	Viable semisterile	Viable but nonreproductive	Viable fertile
Lethal				
• Most monosomics • Trisomics • Polyploids	• XO • XXY	Balanced translocations	• Trisomy 13 • Trisomy 18 • Trisomy 21	• XXX • XYY

Figure 6.7 **Natural selection** operates in human populations to ensure the reduced survival and minimal reproductive potential of individuals with major chromosomal abnormalities.
Source: Data from Chandley, 1981.

largely through pregnancy loss and by sterility or special forms of meiosis that eliminate chromosome complements with extra chromosomes. Nature has created a great barrier to the perpetuation of chromosomally abnormal offspring. Natural selection is not perfect, however. Some chromosomally abnormal fetuses escape nature's screening mechanism and survive to term. An example of this is the Down syndrome infant. About 80 percent of Down infants are aborted spontaneously; only 20 percent are liveborn. Scientists, ethicists, and parents grapple with the question of whether or not we should assume the responsibility for ameliorating nature's "shortcoming" by further limiting the opportunities for the survival of major chromosomal anomalies.

7

SELECTION IN ACTION

The process of evolution is slow and continuous. We have seen that many generations of persistent selection are required to drastically reduce the frequency of an unfavorable mutant gene in a population. Likewise, it generally takes an inordinately long period of time for a new favorable mutant gene to replace its allele throughout a large population. Yet, we have encountered situations in nature in which a favorable mutation has spread through a population in a comparatively short span of years. We shall look at some outstanding examples in which we have actually observed evolution in progress.

INDUSTRIAL MELANISM

One of the most spectacular evolutionary changes witnessed by us has been the emergence and predominance in modern times of dark, or *melanic*, varieties of moths in the industrial areas of England and continental Europe. Slightly more than a century ago dark-colored moths were exceptional. The typical moth in the early 1800s had a light color pattern, which blended with the light coloration of tree trunks on which the moths

alighted. But then the industrial revolution intervened to alter materially the character of the countryside. As soot and other industrial wastes poured over rural areas, the vegetation became increasingly coated and darkened by black smoke particles. In areas heavily contaminated with soot, the formerly abundant light-colored moths were gradually supplanted by the darker varieties. This dramatic change in the coloration of moths has been termed *industrial melanism.* At least 70 species of moths in England have been so affected by the human disturbance of the environment.

During the past two decades, several scientists, particularly E.B. Ford and H. B. D. Kettlewell at the University of Oxford, have analyzed the phenomenon of industrial melanism. Kettlewell photographed the light and dark forms of the peppered moth, *Biston betularia,* against two different backgrounds (fig. 7.1). The light variety is concealed and the dark form is clearly visible when the moths rest on a light lichen-coated trunk of an oak tree in an unpolluted rural district. Against a sooty black oak trunk, the light form is conspicuous and the dark form is well camouflaged. Records of the dark form of the peppered moth date back to 1848,

Figure 7.1 **Dark and light forms** of the peppered moth *(Biston betularia)* clinging to a soot-blackened oak tree in Birmingham, England *(right)* and to a light, lichen-coated oak tree in an unpolluted region *(left)*. Arrow points to a barely visible light peppered moth.

Experimental breeding tests have demonstrated that the two varieties differ principally by a single gene, with the dark variant dominant to the light one. The dominant mutant allele was initially disadvantageous. However, as an indirect consequence of industrialization, the mutant allele became favored by natural selection and spread rapidly in populations in a comparatively short period of time. In unpolluted or nonindustrial areas in western England and northern Scotland, the dominant mutant allele does not confer an advantage on its bearers and the light recessive moth remains the prevalent type.

One of the many impressive features of Kettlewell's studies lies in the unequivocal identification of the selecting agent. Selection, we may recall, has been defined as differential reproduction. The act of selection in itself does not reveal the factors or agencies that enable one genotype to leave more offspring than another. We may demonstrate the existence of selection, yet remain baffled as to the precise causative agent of selection. We might have reasonably suspected that predatory birds were directly responsible for the differential success of the melanic forms in survival and reproduction, but Kettlewell's laboriously accumulated data provided that all-important, often elusive ingredient: *verifiable evidence.*

If the environment of the peppered moth were to become altered again, natural selection would be expected to favor the light variety again. In the 1950s, the British Parliament passed the Clean Air Act, which decreed, among other things, that factories must switch from soft high-sulfur (sooty) coal to less smoky fuels. The enforcement of this enlightened smoke-abatement law has led to a marked reduction in the amount of soot in the atmosphere. In the 1970s, University of Manchester biologist L. M. Cook and his colleagues reported a small but significant increase in the frequency of the light-colored peppered moth in the Manchester

when its occurrence was reported at Manchester in England. At that time, the dark form comprised less than 1 percent of the population. By 1898, only 50 years later, the dark form had come to dominate the Manchester locale, having attained a remarkably high frequency of occurrence estimated at 95 percent. In fact, the incidence of the melanic type has reached 90 percent or more in most British industrial areas.

The rapid spread of the dark variety of moth is certainly understandable. The dark variants are protectively colored in the smoke-polluted industrial regions. They more easily escape detection by predators, namely, insect-eating birds. Actual films taken by Kettlewell and Niko Tinbergen revealed that birds prey on the moths in a selective manner. That is to say, in woodlands polluted by industry, predatory birds more often capture the conspicuous light-colored moths. In a single day, the numbers of light forms in an industrial area may be pared by as much as one-half by bird predation.

area. This is further substantiation of the action and efficacy of natural selection.

CANCER CELL HETEROGENEITY

Until recently, there has been a tendency to focus almost exclusively on environmental factors or carcinogens in the search for the causes of cancer. It cannot be denied that environmental pollutants, such as hydrocarbons, are involved in the etiology of malignancy. But, there is also the realization that carcinogens damage DNA, and that the roots of cancer reside in the genes. Current information indicates that the development of malignancy is a multi-step process involving sequential changes at the DNA level. The process begins when a cell's DNA is altered by some form of mutation which disrupts the harmonious checks and balances that regulate normal cell growth.

Human tumors are *clonal*—that is, they are derived from a single progenitor cell. However, as the tumor grows, the constituent cells become extremely heterogeneous. The heterogeneity represents the emergence of subpopulations, or subclones, of variant cell types. These subclones differ in chromosome numbers, invasiveness, growth rate, responsiveness to blood-borne antibodies, and ability to *metastasize,* or invade other areas of the body (fig. 7.2). The different cell types do not survive fortuitously. Rather, they have been selected for their favorable attributes in perpetuating the existence of the cancerous growth, much to the detriment of the host. Evidently, the process of selection is not confined only to populations of organisms—selection operates on the cellular level as well!

The subpopulations of cells in a solid tumor are continually changing. New variant subpopulations arise unceasingly, each population attempt-

ing to gain a growth advantage over the earlier population. The numbers of chromosomes become altered with time; in fact, chromosomal changes are almost a rule in cancer lineages. Most subpopulations are eliminated because of metabolic disadvantage or immunologic destruction. Those that prevail, or are selectively favored, have the greatest resistance to anticancer drugs.

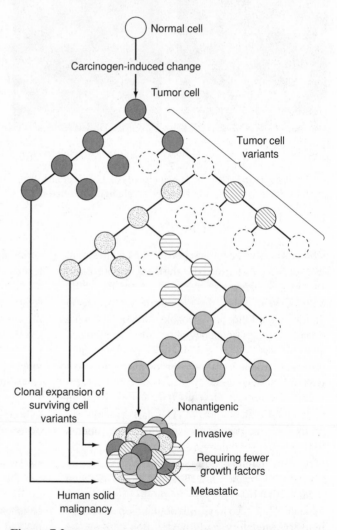

Figure 7.2 Tumor cell heterogeneity. New subpopulations arise continually from the descendants of the original malignant cell. Some subpopulations persist; those that survive and expand differ from each other in certain properties but all are adept at evading host defenses.

The selection process is vividly revealed by the tumor's ability for angiogenesis and metastasis. As a solid tumor grows beyond the diameter of 2 millimeters, its nutritional requirements cannot be satisfied solely by the process of diffusion. Subclones of cell types within the tumor actually elicit the growth of new blood vessels from surrounding host tissue. The new growth of blood vessels is termed *angiogenesis*. The tumor cells themselves synthesize and release the growth factors that stimulate the formation of microvessels. Thus, the progression of cancer is dependent on the selection of potent cells that actually synthesize and secrete their own factors for the growth of blood vessels from pre-existing vessels. Similarly, the metastasis of a tumor to a distant site awaits the emergence of cell types that can direct the tumor to particular organs in the body. The organ chosen is usually one that provides a highly favorable environment for growth of the tumor.

SELECTION FOR RESISTANCE

Penicillin, sulfonamides (sulfa drugs), streptomycin, and other modern antibiotic agents made front-page headlines when first introduced. These wonder drugs were exceptionally effective against certain disease-producing bacteria and contributed immeasurably to the saving of human lives in World War II. However, the effectiveness of these drugs has been reduced by the emergence of resistant strains of bacteria. Medical authorities regard the rise of resistant bacteria as the most serious development in the field of infectious diseases over the past decade. Bacteria now pass on their resistance to antibiotics faster than people spread the infectious bacteria.

Mutations have occurred in bacterial populations that enable the mutant bacterial cells to survive in the presence of the drug. Here again we notice that mutations furnish the source of evolutionary changes and that the fate of the mutant gene is governed by selection. In a normal environment, mutations that confer resistance to a drug are rare or undetected. In an environment changed by the addition of a drug, the drug-resistant mutants are favored and supplant the previously normal bacterial strains.

It might be thought that the mutations conferring resistance are actually caused or induced by the drug. This is not the case. Drug-resistant mutations arise randomly in bacterial cells, irrespective of the presence or absence of the drug. An experiment devised by the geneticist Joshua Lederberg provides evidence that the drug acts as a selecting agent, permitting preexisting mutations to express themselves. As seen in figure 7.3, colonies of bacteria were grown on a streptomycin-free agar medium in a petri plate. When the agar surface of this plate was pressed gently on a piece of sterile velvet, some cells from each bacterial colony clung to the fine fibers of the velvet. The imprinted velvet could now be used to transfer the bacterial colonies onto a second agar plate. More than one replica of the original bacterial growth can be made by pressing several agar plates on the same area of velvet. This ingenious technique has been appropriately called *replica plating*.

In preparing the replicas, Lederberg used agar plates containing streptomycin. On these agar plates, only bacterial colonies resistant to streptomycin grew. In the case depicted in figure 7.3, one colony was resistant. Significantly, this one resistant colony was found in exactly the same position in all replica plates. If mutations arose in response to exposure to a drug, it is hardly to be expected that mutant bacterial colonies would arise in precisely the same site on each occasion. In other words, a haphazard or random distribution of resistant bacterial colonies, without restraint or attention to location in the agar plate, would be expected if the mutations did not already exist in the original bacterial colonies.

Now, we can return to the original plate, as Lederberg did, and test samples of the original bacterial colonies in a test tube for sensitivity or resistance to streptomycin (bottom part of fig. 7.3). It is noteworthy that the bacterial colonies on the original plate had not been previously in contact with the drug. When these original colonies were isolated and tested for resistance to streptomycin, only

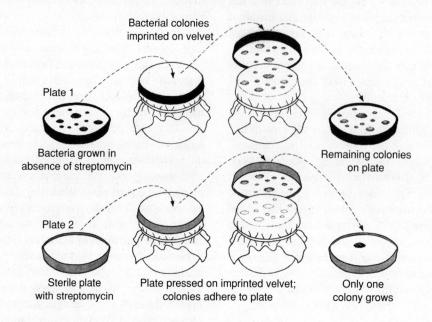

Spontaneous origin of mutation

Bacterial colonies
imprinted on velvet

Plate 1

Bacteria grown in
absence of streptomycin

Remaining colonies
on plate

Plate 2

Sterile plate
with streptomycin

Plate pressed on imprinted velvet;
colonies adhere to plate

Only one
colony grows

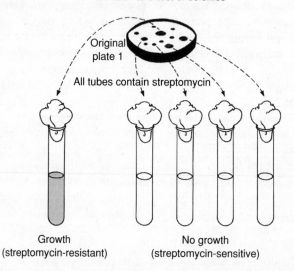

Isolation and test of colonies

Original
plate 1

All tubes contain streptomycin

Growth
(streptomycin-resistant)

No growth
(streptomycin-sensitive)

Figure 7.3 Experiment by Joshua Lederberg revealing that drug-resistant
mutations in bacterial cells had not been induced by the drug but had already been
present prior to exposure of the bacteria to the drug.

one colony proved to be resistant. This one colony occupied a position on the original plate identical with the site of the resistant colony on the replica plates. The experiment demonstrated conclusively that the mutation had not been induced by streptomycin but was already present before exposure to the drug.

MULTIPLE ANTIBIOTIC RESISTANCE

The emergence of antibiotic-resistant bacteria appears deceptively simple. As described above, a spontaneous mutation conferring resistance permits the selective multiplication of mutant bacteria in the presence of a particular antibiotic. The mutation of a gene from sensitivity to resistance would be expected to occur about once in several million divisions, and cells containing the mutant gene would be expected to be resistant to only one particular antibiotic. It would therefore come as an unforeseen surprise to encounter bacteria that *instantaneously* become resistant to *several* antibiotics.

In 1955, a Japanese woman curtailed her visit to Hong Kong because of an unmanageable case of bacillary dysentery, an intestinal inflammation caused by bacteria of the genus *Shigella*. Upon

culture, her intestinal *Shigellae* proved to be unusual. They were found to be resistant simultaneously to four antibiotics commonly used to control bacterial infections—namely, sulfanilamide, streptomycin, chloramphenicol, and tetracycline. Since the late 1950s, there have been innumerable outbreaks of intractable dysentery in Japan, all reflecting the rapid increase of multiply drug-resistant *Shigella dysenteriae*.

A bacterial cell contains a main, large circular strand of DNA and smaller circular pieces of DNA called *plasmids* (fig. 7.4). Plasmids do not have the same sequences of nucleic acid bases as the main ("chromosomal") DNA and replicate independently of the main DNA. A single bacterial cell can harbor as many as 25 plasmids, and each plasmid can carry from 4 to 250 genes, depending on the size of the ring. Plasmids may be viewed as accessory "chromosomes" serving varied functions. One type of plasmid, dubbed the *resistance (R) plasmid,* is clearly adapted to providing the bacterium with substantial resistance to antibiotics. As shown in figure 7.4, four resistance genes, each coding an enzyme that inactivates a specific antibiotic, are linked together on a single R plasmid.

The genetic determinants of an R factor plasmid can be transferred rapidly from one bacterium

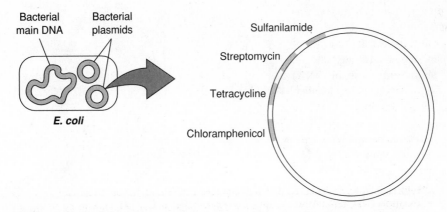

Figure 7.4 A simplified map of an R plasmid, showing four genetic determinants of resistance to four different antibiotics.

to another bacterium by cell-to-cell contact, in a process called *conjugation.* Ironically, R plasmids act like infectious agents as they expeditiously pass copies of themselves to other bacteria during conjugation. In a single conjugation event, several resistance genes can be instantly transferred. New bacterial strains evolved in a matter of months during epidemics of bacterial dysentery in Japan. In 1953, only 0.2 percent of the *Shigellae* isolated from Japanese patients were resistant to antibiotics. By 1965, the level had risen to 58 percent, with virtually all the pathogenic bacteria being resistant to the four aforementioned antibiotics.

Many of the patients who harbored drug-resistant *Shigellae* also quartered strains of the relatively harmless, common colon bacillus *Escherichia coli,* strains which were equally resistant to the four antibiotics. It is quite possible that the *Shigellae* obtained their R plasmids from *E. coli.* There is now experimental evidence that resistance factors can be transferred indiscriminately among different species—for example, *E. coli* can impart its R plasmids to *Shigella dysenteriae,* as well as to *Salmonella typhosa* (the causative agent of typhoid fever), *Klebsiella pneumoniae* (responsible for a virulent form of pneumonia), and *Pasturella pestus* (the bacterium involved in bubonic plague). The rise of pathogenic bacteria with multiple resistance to antibiotics has become a serious problem in human and veterinary medicine.

Remarkably, the resistance genes of a plasmid can become integrated into the bacterial "chromosome" (main DNA) and move from one plasmid to another. Resistance genes are found in highly mobile elements that molecular geneticists refer to as *transposable elements,* or *transposons.* As seen in figure 7.5, a transposon has an appealing architecture. The resistance gene (for tetracycline, in this example) is flanked on both sides by nucleotide sequences that are complementary to each other. These

repeated sequences are called *insertion sequences,* since they are involved only in the insertion of a gene (resistance gene, in this case) into DNA molecules. The complementary sequences form the stem of a "lollipop" during the insertion of the gene, which resides in the head of the lollipop. Strange as it may seem, we identify the existence and position of a transposon by the lollipop configuration in a DNA preparation. Transposons can move readily between plasmids as well as integrate into bacterial or viral "chromosomes."

The transposable elements are not merely a curiosity of bacteria. Transposable elements have been discovered in yeast, the fruit fly, corn, and primates, including humans. Transposons are frequently spoken of as *jumping genes.* The pioneering studies on jumping genes in corn were performed in the 1940s by Barbara McClintock of Cold Spring Harbor Laboratory in New York. One cannot do justice to the analytical power of the innovative experiments conceived by her. It merits attention that on December 10, 1983, the Nobel prize for medicine or physiology was awarded to Barbara McClintock for her discovery of unstable

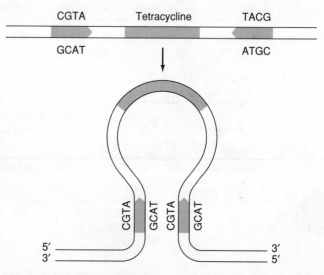

Figure 7.5 **The structure of a transposon** bearing a resistance gene for tetracycline and its conversion into a stem-loop configuration ("lollipop").

jumping genes. It took nearly 40 years for the genetic community to accept McClintock's heterodox genes, since the notion that genes were unstable both in their location and in their function was too disruptive to accept.

There are many mutiples of relatively short sequences of nucleotides in the human genome, which we may call simply *repetitive DNA.* The large number of repetitive nucleotide sequences in humans represents an awesome potential for the jumping of genes from one segment of the DNA molecule to another. The short repeats can promote their own movement from one location in the DNA molecule to another. These jumping repeats tend to inactivate genes in their new location or, at least, modify the expression of the genes in which they lodge. A stream of new information implicates transposable elements in several human genetic disorders, notably hemophilia A and type 1 neurofibromatosis. In its usual location, the repetitive piece of DNA lacks the capacity to direct the production of a protein. Its insertion, however, into a gene that does code for a protein leads to an abnormal protein product and the expression of a genetic defect. In large measure, jumping genes in humans act as mutagenic agents.

"Prudent Use" of Antibiotics in Human Medicine

Many human advocacy groups, both private and public, have clamored for a comprehensive program of monitoring the use of antibiotics to forestall the seemingly endless emergence of antibiotic-resistant bacteria. The conventional wisdom is that an evolved antibiotic-resistant bacterial colony would become sensitive again in an environment restored to normal. However, studies by Bruce Levin and his associates at Emory University have demonstrated, using *E. coli* strains, that mutations at different sites compensate for the presumed loss of fitness associated with resistance genes in a normal environment and create a genetic background in which drug-sensitive genes actually are at a

marked selective disadvantage! Stated another way, the emergence of so-called *compensatory mutations* precludes the evolved resistant lineages from reverting to drug sensitivity in a normal environment. If resistant lineages persist in a normal environment, this casts serious doubt as to whether the prudent use of antibiotics advocated by health officials, admirable in itself, can arrest the torrent rate of bacterial resistance to antibiotics.

Recombinant DNA

The painstaking studies of bacterial plasmids by a host of investigators have paved the way for the imaginative use of plasmids as carriers of foreign DNA. The plasmids can be manipulated in such a fashion that they can carry genes from very distantly related organisms, including frogs, rabbits, and humans. This sophisticated technology involves the use of a special class of enzymes called *restriction endonucleases.* These enzymes act as chemical scissors; they cut a DNA strand at precise points.

Restriction endonucleases occur in bacteria where they function, it seems, as primitive immune systems, destroying the DNA of invading viruses. There are scores of these enzymes in existence, around 2,500 exhibiting a variety of different specificities. Each of these enzymes recognizes a particular short nucleotide sequence in any DNA exposed to it and snips the DNA at that point. If the sequence occurs more than once, as it generally does, the DNA molecules are cut in several fragments. The endonucleases were originally discovered in bacteria about 25 years ago by Werner Aber of the University of Basel (Switzerland) and by the team of Daniel Nathans and Hamilton D. Smith at Johns Hopkins University, for which achievement the three received the 1978 Nobel Prize for Medicine. Among the first restriction enzymes purified were *Eco*RI and *Eco*RV, from *Escherichia coli,* and *Hind*II and *Hind*III, from *Haemophilus influenzae.* Some restriction enzymes and the base sequences they cleave are listed in table 7.1. As an example, *Eco*RI will cut a DNA molecule from any source at any point at which the nucleotide sequence GAATTC is encountered.

Restriction enzymes are species nonspecific— that is, enzymes of the same specificity occur in different bacterial species. However, each bacterial strain contains only one to a few different restriction enzymes. The genes for restriction enzymes, like the genes for drug resistance, shuffle between bacterial strains. Natural selection for variety continually occurs. There are never-ending cycles of adaptive, outmaneuvering viral variants that evade the bacteria's restriction enzymes. This leads to a new generation of bacteria that, by acquisition or mutation, have new enzymes that restrict the viral variants. This, in turn, selects for the next generation of viruses now able to evade the new enzymes. The bacteria also synthesize "modification" enzymes that protect their *own* DNA from cleavage. These enzymes recognize the same DNA sequence as the restriction enzyme they accompany, but instead of cleaving the sequence, they disguise it by methylating one of the bases in each DNA strand.

Several thousand human genes have already been isolated and cloned. The isolated human gene can be spliced into a plasmid that has been experimentally fragmented at specific points by a restriction endonuclease (fig. 7.6). The outcome is a new

TABLE 7.1	Some Restriction Enzymes and the Base Sequences They Cleave

Bacterial source	Enzyme	Sequence*
Escherichia coli	*Eco*RV	5'...G A T|A T C ...3' 3'...C T A|T A G ...5'
Escherichia coli	*Eco*RI	5'...G|A A T T C ...3' 3'...C T T A A|G ...5'
Haemophilus haemolyticus	*Hha*I	5'...G C G|C ...3' 3'...C|G C G ...5'
Providencia stuarti	*Pst*I	5'...C T G C A|G ...3' 3'...G|A C G T C ...5'

* The upstream portion of a gene sequence is referred to as the 5' end; the downstream portion, the 3' end.

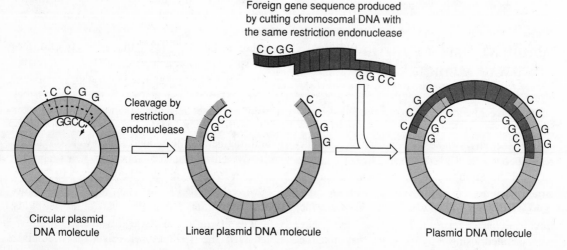

Foreign gene sequence produced by cutting chromosomal DNA with the same restriction endonuclease

C C G G

G G C C

Circular plasmid DNA molecule

Cleavage by restriction endonuclease

Linear plasmid DNA molecule

Plasmid DNA molecule

Figure 7.6 A recombinant DNA molecule, containing both plasmid (bacterial) DNA and human DNA. The circular plasmid DNA is cut by a restriction endonuclease that leaves cohesive single-stranded ends. A foreign (human) DNA sequence is generated by the same restriction endonuclease and, accordingly, has complementary cohesive ends that permit base-pair interactions.

DNA circle, or *recombinant DNA molecule,* which contains both plasmid DNA and human DNA. This newly formed recombinant DNA molecule, when reintroduced into a bacterium, can replicate precisely and be passed on to daughter bacterial cells for many generations. Given that a bacterial cell can divide every 30 to 40 minutes, a single bacterium, under ideal growing conditions, can produce almost 70 billion new cells in 24 hours. At the same time, billions of copies of the product of the human gene become available!

This ingenious technique has permitted the large-scale transmission of genes that code for commercially indispensable molecules. As an admirable example, the insertion of the human insulin-producing gene into *Escherichia coli* has made possible the production of insulin in large quantities. Another notable example is the production of human growth hormone. In this instance, the genes coding for the polypeptide chains of the growth hormone were actually synthesized chemically in the laboratory. The synthetically created genes were incorporated into *E. coli* and efficient synthesis of human growth hormone was achieved (fig. 7.7). In addition to human insulin and growth hormone, other valuable proteins, such as ovalbumin and interferon, have been produced by recombinant DNA technology.

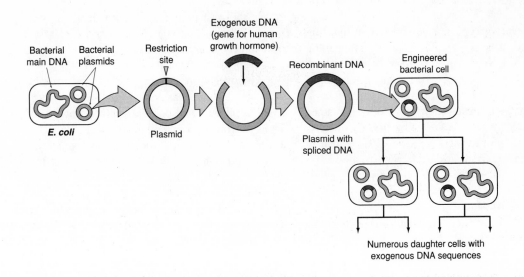

Figure 7.7 **Production** of virtually limitless quantities of an exogenous gene product (human growth factor, for example) is accomplished by inserting the exogenous gene in a bacterial plasmid.

8

BALANCED POLYMORPHISM

The concepts presented in the preceding chapter have led us to believe that selection operates at all times to reduce the frequency of a detrimental gene to a low equilibrium level. This view is not entirely accurate. We are aware of genes with deleterious effects that occur at fairly high frequencies in natural populations. A striking instance is the high incidence in certain human populations of a mutant gene that causes a curious and life-threatening form of blood cell destruction, known as *sickle-cell anemia*. It might be presumed that this harmful gene is maintained at a high frequency by an exceptionally high mutation rate. There is, however, no evidence to indicate that the sickle-cell gene is unusually mutable. We now know that the maintenance of deleterious genes at unexpectedly high frequencies involves a unique, but not uncommon, selective mechanism, which results in a type of population structure known as *balanced polymorphism*.

SICKLE-CELL ANEMIA

This disease was discovered by the American physician James B. Herrick, who in 1904 made an office examination of an anemic African American male residing in Chicago. The patient's blood examined under the microscope showed the presence of numerous crescent-shaped erythrocytes. The peculiarly twisted appearance of the red blood cell is shown in figure 8.1. The patient was kept under observation for six years, during which time he displayed many of the distressing symptoms we now recognize as typical of the disease.

The bizarre-shaped red cells in the form of a sickle blade cannot easily negotiate the thin spaces of the capillaries and cause miniature "log jams." The clogging of small blood vessels can occur anywhere in the body, denying vital oxygen to the tissues and bringing about the painful crises that are so characteristic of the disease. Life expectancy is reduced—half the victims succumb before the age of 20 and most do not survive beyond 40. The clinical picture, however, is quite variable, with some patients having only few crises and mild pains for many years, while others become severely disabled or die at an early age.

Sickle-cell anemia occurs predominantly among African Americans. The incidence at birth of the disabling sickle-cell anemia in the United States is estimated at 2 per 1,000 infants. Not until

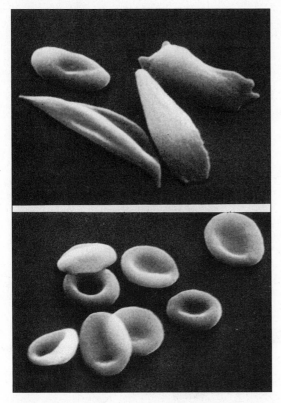

Figure 8.1 Peculiarly twisted shape of red blood cells of an individual suffering from sickle-cell anemia *(top),* contrasted with the spherical appearance of normal red blood cells *(bottom).* (World Wide Photos.)

be recessively determined. However, although heterozygous individuals are, on the whole, normal, even the red cells of the heterozygotes can undergo sickling under certain circumstances, producing clinical manifestations. As an instance, heterozygous carriers have been known to experience acute abdominal pains at high altitudes in unpressurized planes. The lowering of the oxygen tension, with an increased amount of oxygen released from chemical combination, is sufficient to induce sickling. The abdominal pains can be traced to the packing of sickled erythrocytes in the small capillaries of the spleen.

According to some authors, since the detrimental gene can express itself in the heterozygous state (by producing a positive sickling test under certain circumstances), the gene should be considered dominant. This reveals that "dominance" and "recessiveness" are somewhat arbitrary concepts that depend on one's point of view. Indeed, from a molecular standpoint, the relation between the normal and variant allele in this case may best be described as *co-dominant.* This means that at the molecular level neither allele masks the expression of the other. As we shall see below, the heterozygote produces both normal and abnormal hemoglobin. *There is no blending of inheritance at the molecular level.*

after World War II was its hereditary basis elucidated. The irregularity was shown in the late 1940s by James V. Neel of the University of Michigan to be inherited as a simple Mendelian character. The sickle-cell anemic patient inherits two variant (sickling) alleles, one from each parent (fig 8.2). Individuals with one normal and one variant allele are generally healthy but are carriers—they are said to have *sickle-cell trait.* If two heterozygous carriers marry, the chances are one in four that a child will have sickle-cell anemia, and one in two that a child will be a carrier (fig. 8.2).

Since the homozygous state of the detrimental allele is required for the overt expression of the disease, sickle-cell anemia may be considered to

SICKLE-CELL HEMOGLOBIN

In 1949, the late distinguished chemist and Nobel laureate Linus Pauling and his co-workers made the important discovery that the detrimental sickling allele alters the configuration of the hemoglobin molecule. Pauling used the then relatively new technique of electrophoresis, which characterizes proteins according to the manner in which they move in an electric field. The hemoglobin molecule travels toward the positive pole. Hemoglobin from a sickled cell differs in speed of migration from normal hemoglobin; it moves more slowly than the normal molecule (fig. 8.3). The variant allele thus functions differently from the normal allele, and, in fact, acts independently of its normal

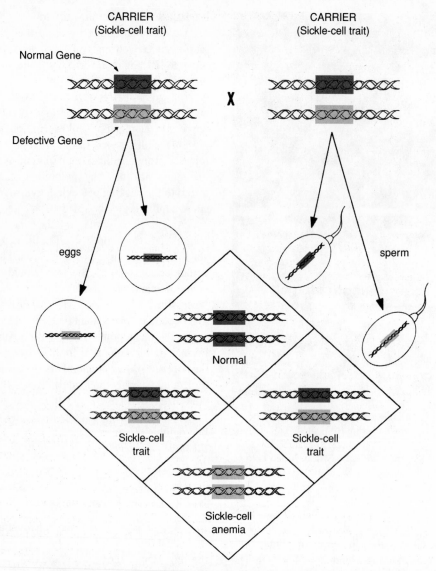

Figure 8.2 Types of children that can result from a marriage of two heterozygous carriers of sickle-cell anemia. Individuals homozygous for the variant gene suffer from sickle-cell anemia: the benign heterozygous state is referred to as the sickle-cell trait.

partner allele. As a result, the heterozygote does not produce an intermediate product, but instead produces both kinds of hemoglobin in nearly equal quantities—the normal type (designated *hemoglobin A,* or *Hb A*) and the sickle-cell anemic variety *(Hb S).* The dual electrophoretic pattern of hemo-

globin from a heterozygous individual can actually be duplicated experimentally by mechanically mixing the hemoglobin taken from blood cells of a normal person and a sickle-cell anemic patient. The mixed solution separates in an electric field into the same two hemoglobin components as

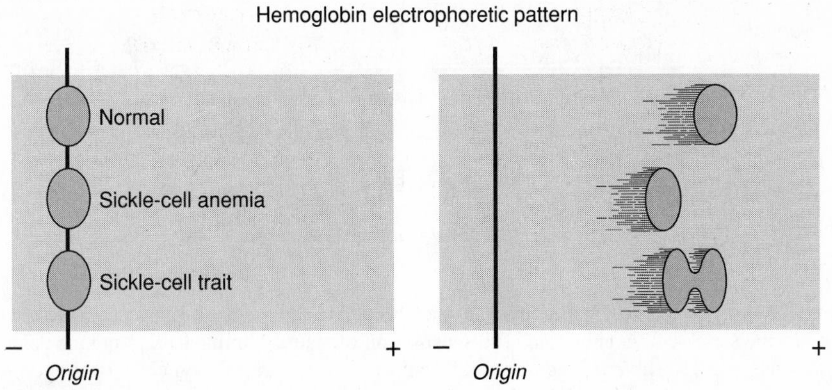

Hemoglobin electrophoretic pattern

Figure 8.3 Electrophoretic patterns of hemoglobins. The hemoglobin of the
heterozygous person with sickle-cell trait is not intermediate in character, but is
composed instead of approximately equal proportions of the normal hemoglobin and
the sickle-cell anemia variety. [Based on studies by Linus Pauling, in Pauling, Itano,
Singer, and Wells, "Sickle-Cell Anemia: A Molecular Disease," *Science* 110
(25 November 1949): 543–548.]

those characteristic of a heterozygous person with
sickle-cell trait.

Further chemical analysis revealed that sickle-
cell hemoglobin differed only slightly in its chemical constitution from normal hemoglobin. As mentioned earlier (chapter 4), a glutamic acid residue at
a specific site in normal hemoglobin is replaced by
valine in sickle-cell hemoglobin (fig. 8.4). The
substitution of valine for glutamic acid can be
accounted for by an alteration of a single base
change of the triplet of DNA that specifies the
amino acid, as depicted in figure 8.5. If the DNA
triplet responsible for transcription in normal
hemoglobin *(Hb A)* is CTT, then a single base
change to CAT would alter the codon of messenger RNA to GUA. Accordingly, the amino acid
specified by the codon GUA would be valine. The
substitution of a glutamic acid by valine causes the
abnormal hemoglobin molecule *(Hb S)* to aggre-
gate or polymerize into strands. The strands are
laid down to form cable-like fibers. As greater
numbers of fibers accumulate, the large aggregates
or polymers of linearly arranged fibers attain sufficient length and rigidity to distort the cell membrane into an odd, crescent shape.

THEORY OF BALANCED POLYMORPHISM

Since many persons with sickle-cell anemia do not
survive to reproductive age, it might be expected
that the detrimental allele would pass rapidly from
existence. Each failure of the homozygous anemic
individual to transmit his or her alleles would
result each time in the loss of two aberrant alleles
from the population. And yet, the sickle-cell allele
reaches remarkably high frequencies in the tropical zone of Africa. In several African populations,
20 percent or more of the individuals have the
sickle-cell trait, and frequencies as high as 40 percent have been reported for some African tribes.
The sickle-cell trait is not confined to the African
continent; it has been found in Sicily and Greece,
and in parts of the Near East. What can account for
the high incidence of the sickle-cell allele, particularly in light of its detrimental action?

For simplicity in the presentation of the population dynamics, we will consider the sickling
allele to be recessive. Accordingly, we can sym-
bolize the normal allele as *A* and its allele for sickling as *a*. Now, the explanation for the high level

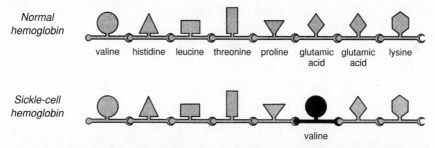

valine   histidine   leucine   threonine   proline   glutamic   glutamic   lysine
                                                          acid   acid

valine

Figure 8.4 **Amino acid sequences** in a small section of the normal hemoglobin molecule and of the sickle-cell hemoglobin. The substitution of a single amino acid, glutamic acid by valine, is responsible for the abnormal sickling of human red blood cells. (Based on studies by Vernon Ingram.)

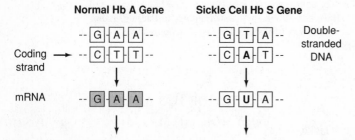

Figure 8.5 **The CTT in the sixth triplet** of the coding strand of the normal gene changes to CAT. This leads to a change from GAA to GUA in the sixth codon of the beta-globulin mRNA of sickle cells. This, in turn, results in the insertion of a valine in the sixth amino acid position of sickle-cell beta globulin, where a glumatic acid normally is present.

of the deleterious sickle-cell allele is to be found in the possibility that the heterozygote *(Aa)* is superior in fitness to *both* homozygotes (AA and aa). In other words, selection favors the heterozygote, and both types of homozygotes are relatively at a disadvantage. Let us examine the theory behind this form of selection.

Figure 8.6 illustrates the theory. The classical case of selection discussed in preceding chapters is portrayed in the first part of the figure. In this case, when AA and Aa individuals are equal in reproductive fitness, and the aa genotype is completely selected against, the recessive allele will be eliminated. Barring recurring mutations, only A genes will ultimately be present in the population.

Let us assume in another situation (case II in fig. 8.6) that both homozygotes (AA as well as aa) are incapable of leaving surviving progeny. The only effective members in the population are the heterozygotes *(Aa)*. Obviously, the frequency of each allele, A and a, will remain at a constant level of 50 percent. The inviability of both homozygous types probably never exists in nature, but the scheme does reveal how a stable relation of two alleles at high frequencies is possible. If, as in case III, the AA genotype leaves only half as many progeny as the heterozygote, and the recessive homozygote is once again inviable, it is apparent that more A alleles are transmitted each generation than a alleles. Eventually, however, the A allele will reach an equilibrium point at 0.67. Here, it may be noted that although the recessive homozygote is lethal, the frequency of the recessive allele *(a)* is maintained at 0.33. In the last illustrative example (case IV), the recessive homozygote is not as disadvantageous as the dominant homozygote, but both are less reproductively fit than the heterozygote. Here also, both alleles remain at relatively high frequencies in the population. Indeed, the recessive allele *(a)* will constitute 67 percent of the gene pool.

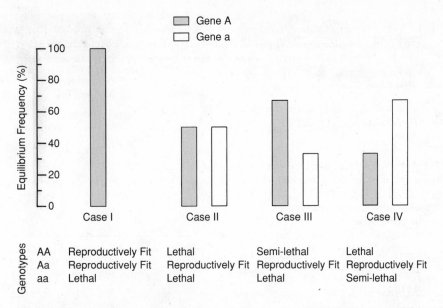

Figure 8.6 **Equilibrium frequencies** of two alleles *(A* and *a)* under different conditions of selection (ignoring mutation). In contrast to case I (complete selection and total elimination of *a)*, the recessive gene can be retained at appreciable frequencies in a population when the heterozygote *(Aa)* is superior in reproductive fitness to *both* homozygotes (cases II, III, and IV, showing different relative fitnesses of the two homozygotes, *AA,* and *aa).*

We have thus illustrated in simplified form the selective forces that serve to maintain two alleles at appreciable frequencies in a population. This phenomenon is known as *balanced polymorphism.* The loss of a deleterious recessive allele through deaths of the homozygotes is balanced by the gain resulting from the larger numbers of offspring produced by the favored heterozygotes.

Balanced polymorphism results in a stable equilibrium of the two alleles; their equilibrium frequencies are determined by the relative fitness of the two homozygotes. If the fitness of *Aa* is set equal to 1, of *AA* equal to $(1 - s_1)$, and of *aa* equal to $(1 - s_2)$, where s denotes the selection coefficient, then the frequency of the recessive allele q_a at equilibrium can be calculated by the following formula:

$$q_a = \frac{s_1}{s_1 + s_2}.$$

An example of the use of the formula is shown in table 8.1. When the relative fitness of the recessive homozygote is only 0.10 and that of the dominant homozygote is 0.85, the obviously harmful recessive allele still reaches a high stable frequency of about 14 percent.

EXPERIMENTAL VERIFICATION OF BALANCED POLYMORPHISM

A simple experiment, using the fruit fly, *Drosophila melanogaster,* can be performed to demonstrate that an obviously lethal gene can be maintained at a stable, relatively high frequency in a population. The geneticist P.M. Sheppard of the University of Liverpool in England introduced into a breeding cage a population of flies of which 86 percent were normal and 14 percent carried the mutant allele *stubble,* which affects the bristles of

TABLE 8.1	Calculation of Equilibrium Frequency of Recessive Gene in a Balanced Polymorphic State

1. Heterozygote *(Aa)* is superior in fitness to both homozygotes *(AA* and *aa)* and is assigned a value of 1.

	AA	Aa	aa
Fitness (f)*	$1 - s_1$	1	$1 - s_2$

2. Equilibrium value of recessive gene *a:*

$$q_a = \frac{s_1}{s_1 + s_2}$$

3. Example: If fitness *(f)* of *AA* = 0.85, then selection coefficient (s_1) = 0.15

If fitness *(f)* of *aa* = 0.10, then selection coefficient (s_2) = 0.90

$$q_a = \frac{0.15}{0.15 + 0.90} = \frac{0.15}{1.05} = 0.14$$

*Fitness (f), another measure of the reproductive efficiency of a particular genotype, is the converse of the selection coefficient (s).

TABLE 8.2	Experimental Balanced Polymorphism

Generation	Percentage of Stubble Flies	Frequency of Stubble Gene
1	14.3	0.0715
2	33.7	0.1685
3	57.6	0.2880
4	63.2	0.3160
5	69.1	0.3455
6	73.5	0.3675
7	72.9	0.3645
8	73.4	0.3670
9	72.9	0.3645

From P. M. Sheppard, *Natural Selection and Heredity.* New York: Philosophical Library, 1959.

the fly. The stubble allele is lethal when homozygous; hence, all stubble individuals are heterozygous. Ordinarily, the heterozygous fly does not have any reproductive advantage over the normal homozygote. Sheppard, however, created a situation whereby the heterozygote would be favored: he removed 60 percent of the normal flies from the population each generation. Consequently, the heterozygote was rendered superior in fitness to either homozygote by virtue of the natural lethality of the stubble homozygote and the enforced reproductive incapacity of many of the normal homozygotes. The results of the experiment are recorded in table 8.2.

It should be noted that the frequency of the stubble allele increased in the early generations, but then became stabilized at a level of about 0.365. The equilibrium level is reached when as many stubble alleles are lost from the population, through death of the stubble homozygotes, as are gained as a result of the reproductive advantage of the heterozygote. Although the stubble allele is lethal, the population, under the constant condi-

tions of the experiment, remained at a stable state with a high number of heterozygotes.

The frequency of the stubble allele will fall rapidly when the usual reproductive potential of the normal homozygote is restored. When the normal homozygotes and heterozygotes are equal in reproductive fitness (refer to case I, fig. 8.6), the normal allele will supplant the detrimental stubble allele in the population. With these considerations in mind, we shall return to our discussion of sickle-cell anemia.

SUPERIORITY OF THE HETEROZYGOTES

The high frequency of the sickle-cell allele in certain African populations can be explained by assuming that the heterozygotes (individuals with sickle-cell trait) have a selective advantage over the normal heterozygotes. What might be the nature of the advantage? Field work undertaken in Africa by the British geneticist Anthony Allison revealed that the incidence of the sickle-cell trait is high in regions where malignant subtertian malaria caused by the parasite *Plasmodium falciparum* is hyperendemic—that is, transmission of the infection occurs throughout most of the year. Thus, the population is almost constantly reinfected with malaria. Under such circumstances, relative immunity to falciparum malaria would be most beneficial.

Allison examined blood from African children and found that carriers of the sickle-cell trait were relatively resistant to infection with *Plasmodium falciparum.* The heterozygous carriers were infected less often with the parasite than the homozygous dominant nonsicklers. Moreover, among those heterozygotes that were infected, the incidence of severe, or fatal, attacks of malaria was strikingly low. The evidence is strong that the sickle-cell allele affords young children some degree of protection against malarial infection. Hence, in areas where malaria is common, children possessing the sickle-cell trait will tend to survive malaria, and are more likely to pass on their genes to the next generation. The heterozygotes *(Aa)* are thus superior in fitness to both homozygotes, which are likely to succumb from either anemia on the one hand *(aa)* or malaria on the other *(AA).*

The spread of the sickling allele was greatly enhanced by the development of agriculture in Africa. The clearing of the forest in the preparation of ground for cultivation provided new breeding areas for the mosquito *Anopheles gambiae,* the vector, or carrier, of the malaria plasmodium. The spread of malaria has been responsible for the spread of the selective advantage of the sickle-cell allele, which in the heterozygous state imparts resistance to malaria. In essence, the selective advantage of the heterozygote tends to increase in direct proportion to the amount of malaria present in a given area. Hunting populations in Africa show a very low incidence of malaria and an equally low frequency of the sickling allele. The Pygmies of the Ituri Forest constitute a good example.

Communities in Africa with the greatest reliance on agriculture (rather than on hunting or animal husbandry) tend to have the highest frequencies of the sickle-cell trait. A high incidence of the sickle-cell trait in an intensely malarious

environment has the consequence of reducing the number of individuals capable of being infected by the malarial parasite and, accordingly, of lowering the mortality from such infections. More human energy, or greater "manpower," is thus made available for raising and harvesting crops. Ironically, then, the sickle-cell trait carries with it the beneficent effect of enabling tribes to develop and maintain an agricultural culture rather than adhere to a hunting or pastoral existence. This is a curious but striking instance of the interplay of biological change and socioeconomic adaptation.

In a similar finding, evidence exists that persons whose red blood cells are deficient in the enzyme G6PD (glucose-6-phosphate-dehydrogenase) are less likely to be affected by malaria. Data also show a strong correlation between the incidence of malaria and the frequency of thalassemia (another type of anemia) in the Italian peninsula and in Sardinia. In all these situations, it is as if one inherits something that is bad (hemolytic disease) to afford protection against something that is worse (malaria). Malaria apparently has had a profound influence on human events.

RELAXED SELECTION

We should expect the frequency of the sickle-cell allele to be low in malaria-free areas, where the selective advantage of the heterozygote would be removed. We do find that the lowest frequencies of the sickle-cell allele occur consistently in regions free of malaria. The frequency of the sickle-cell allele has fallen to relatively low levels in the African American populations of the United States. The frequency of the sickle-cell trait among African Americans is currently about 9 percent, corresponding to an allele frequency of 0.045. This places the frequency of the recessive homozygote (sickle-cell anemia) at 0.002, or 2 per 1,000 individuals.

The relaxation of selection in this case is eugenic; that is, the changes occur in the desired direction. The disappearance of malaria disrupts the balanced polymorphic state, and the allele for sickle-cell anemia begins to decline at a slow rate. Table 8.3 presents computations of the gene frequencies of the heterozygotes and the recessive homozygotes, starting with the frequency of the carrier at 18 percent. It may be noted that 30 generations are required for the incidence of the heterozygous carrier to be depressed from 18.0 to 6.5 percent. At present, medical researchers are attempting to find a cure for sickle-cell anemia. If sickle-cell anemic individuals could be completely cured, then obviously the selective process would be thwarted and the sickle-cell allele would no longer decline in frequency.

TAY-SACHS DISEASE

Tay-Sachs disease is a recessively inherited condition that is untreatable and fatal. The basic defect is the abnormal accumulation in the brain cells of a fatty substance (specifically, ganglioside GM_2) as a consequence of the absence of a particular enzyme (hexosaminidase A). The storage of massive amounts of the lipid in the brain leads to profound mental and motor deterioration. Affected children appear normal and healthy at birth. Within six months, however, motor weaknesses become obvious as the muscles twitch and the infant experiences periodic convulsions. The muscles deteriorate until the infant becomes completely helpless, unable to sit up or stand. At the age of one year, the child lies still in the crib. The child

becomes mentally retarded, progressively blind, and finally paralyzed. The disease exacts its lethal toll by the age of three to four years. There are no known survivors and no cure.

A feature of special interest is that 9 out of 10 affected children are of Jewish heritage. It is especially common among the Ashkenazi Jews of northeastern European origin, particularly from provinces in Lithuania and Poland. In the United States, Tay-Sachs disease is about 100 times more prevalent among the Ashkenazi Jews than among other Jewish (Sephardi) groups and non-Jewish populations. It is estimated that 1 of 40 Jewish persons is heterozygous, whereas only 1 of 380 non-Jewish persons is a heterozygous carrier. If the high incidence of heterozygotes is maintained by mutation alone, then an extraordinarily high mutation rate of the detrimental recessive allele would have to be postulated. However, some reproductive advantage for the heterozygote carrier would seem the most plausible explanation.

It has been suggested that the Ashkenazi Jews, who have lived for many generations in the urban ghettos in Poland and the Baltic states, have been exposed to different selective pressures than other Jewish groups (for example, those who have lived in countries around the Mediterranean and Near East). The densely populated urban ghettos may have experienced repeated outbreaks of infectious diseases. In 1972, the geneticist Ntinos Myrianthopoulos of the National Institutes of

TABLE 8.3	Selection Relaxation Against Sickle-Cell Anemia (in Nonmalarial Areas)
	Conditions: Initial fitness of recessive homozygote: 0.33
	Initial gene frequency of sickling gene: 0.10
	Initial frequency of heterozygote: 0.18

Generation	% Heterozygote (Sickle-Cell Trait)	% Recessive Homozygote (Sickle-Cell Anemia)
0	18.0	1.0
1	17.0	0.88
3	15.3	0.70
10	11.4	0.37
30	6.5	0.11
100	2.6	0.02

Health in Bethesda, Maryland, presented data that show that pulmonary tuberculosis is virtually absent among grandparents of children afflicted with Tay-Sachs disease, although the incidence of Jewish tuberculosis patients from eastern Europe is relatively high. The findings suggest that the heterozygous carrier of Tay-Sachs disease is resistant to pulmonary tuberculosis in regions where this contagious disease is prevalent.

In terms of theoretical expectations, one can calculate (in the absence of genetic screening) that 50 children will be born with the disorder each year in the United States, of whom 45 will be of Ashkenazi Jewish origin. Yet, the number of actual cases of affected newborns is far less than expected. Indeed, within the last ten years, the incidence of births of children afflicted with Tay-Sachs disease has been so dramatically reduced that less than a dozen cases are now known in the United States.

The dramatic reductions in births of affected infants reflects a voluntary, widespread adult screening program initiated in 1971 among the Jewish population with the objective of detecting carriers of the fatal recessive gene. The screening for heterozygotes has been followed by the intrauterine monitoring of the fetus in instances where both parents are heterozygous carriers. In most cases, when parents have discovered through screening and subsequent prenatal diagnosis that they have conceived an affected child, they have opted to terminate the pregnancy. The decision is generally not thought of as a choice between life and death for the affected fetus. The choice is between prenatal death and a lingering, painful, postnatal death. At issue is not the saving of the life of an affected child, but the prevention of undeniable suffering.

CYSTIC FIBROSIS

Cystic fibrosis is not a rare genetic condition; one child in 2,500 is born with the recessive disorder. This means that approximately 1,600 affected newborns may be expected each year in the United States. Virtually all patients with cystic fibrosis are Caucasians; African Americans and Orientals are rarely afflicted. In fact, cystic fibrosis represents the most common autosomal recessive disease in Caucasians.

Cystic fibrosis is a multi-system disease, affecting many parts of the body. Specifically, cystic fibrosis is identified by a triad of abnormal conditions—a highly elevated concentration of chloride in sweat, pancreatic insufficiency, and chronic obstructive pulmonary disease. Almost equally consistent is azoospermia (lack of sperm in the semen), seen in more than 95 percent of men with cystic fibrosis. Almost universally, the sperm ducts atrophy, or degenerate, as a consequence of prolonged blockage by thick mucus secretions. The female patient does not have a comparable anatomical abnormality. However, her fertility is below normal, as the thick desiccated cervical mucus tends to be inhospitable to the sperm cells. One in every three affected females is married, and among affected women who become pregnant, there is an increased incidence of prematurity and perinatal deaths.

The frequency of heterozygous carriers is many times greater than the frequency of affected homozygous individuals. Calculations indicate that *1 of 25* persons is a carrier of cystic fibrosis. Stated another way, there are 100 times as many genetic carriers of cystic fibrosis as there are individuals afflicted with this defect. A reproductive advantage for the heterozygote *(Aa)* over the normal homozygote *(AA)* would account for the high frequency of the disabling recessive allele for cystic fibrosis in Caucasian populations. The hypothesis currently favored is that heterozygous carriers are more resistant to infantile gastroenteritis (inflammation of the stomach and intestines) than noncarriers. The basic defect expressed in cystic fibrosis is a failure of transport of chloride ions across cell membranes. Ironically, the same chloride ion channels are involved in another devastating disease with high mortality, *cholera*. This disease is marked by excessive diarrhea and vomiting, associated with the profuse passage of chloride ions through channels in the cells lining the intestinal tract into the lumen of the digestive tract. In a curious manner, it may be that the heterozygous carrier of cystic fibrosis is resistant to cholera in regions where this contagious, infectious disease is, or once was, prevalent.

In 1989, the gene responsible for cystic fibrosis was identified on the long arm of chromosome 7. Knowledge of the precise location of the mutant allele for cystic fibrosis has permitted the development of a diagnostic test for family members at risk. The thrust for prenatal screening may well be a two-edged sword. The survival of patients with cystic fibrosis has improved dramatically in the last two decades. In the mid-1960s, affected children could not look forward to reaching the age of 10 years, despite the best efforts of the medical profession. Today, this disorder still remains a burden to children, but a limited life span is no longer foreboding or inevitable. Presently, with increased clinical awareness and aggressive therapeutic measures, more than half the babies born with this disorder today can expect to live at least to the late twenties (fig. 8.7). Cystic fibrosis has graduated from the pediatric clinic and has become a disease of adults. The disorder, however, still retains its distressing aspects, and the greater longevity may be viewed as imposing greater rather than lesser burdens on the patients and their families. Given these circumstances, will parents opt for prenatal diagnosis when a cure for cystic fibrosis has yet to be found?

Nearly all young mothers with a cystic fibrosis child wish to become pregnant again *without* the 1 in 4 risk of another affected child. Most mothers want more assurance than mere statistical probabilities. It is not uncommon for parents to profess that they do not wish to expose another child to the symptoms and distresses of the disease. Many parents have been influenced by the feeling of affected older children, many of whom are desperately anxious that a younger brother or sister should not have cystic fibrosis.

SELECTION AGAINST THE HETEROZYGOTE

The mother-child blood incompatibility in humans, *erythroblastosis fetalis,* represents an interesting case of selection *against* the heterozygote. This unique hemolytic disorder is often referred to simply as *Rh disease* because it involves the Rh antigen on the red blood cell. Rh disease was once an

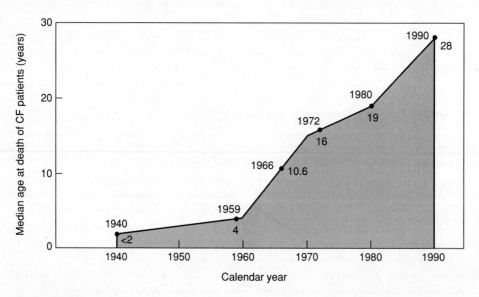

Figure 8.7 Medical advances in recent decades have enhanced the life span of infants affected with cystic fibrosis.

alarming thorn to procreation. The conquest of this curious incompatibility between mother and child is one of the more notable accomplishments of modern medicine.

A relation between the Rh antigen and hemolytic disease was first postulated by Dr. Philip Levine and his colleague Dr. B. E. Stetson in 1939, when they discovered a then unknown antibody in the serum of a woman who had recently delivered a stillborn infant. The antibody was subsequently identified as anti-Rh, which is produced by the mother in response to, and directed against, the Rh antigen of the blood cells of her own fetus. Clinical records reveal that the mothers of affected new-borns lack the Rh antigen (that is, are *Rh-negative*) whereas their husbands and affected infants possess the Rh antigen (that is, are *Rh-positive*).

The Rh antigen in the blood cell is controlled by a dominant allele, designated *R.* An Rh-positive person has the dominant allele, either in the homozygous *(RR)* or heterozygous *(Rr)* state. All Rh-negative individuals carry two recessive alleles *(rr)* and are incapable of producing the Rh antigen. The inheritance on the Rh antigen follows simple Mendelian principles (fig. 8.8). A mother who is Rh-negative *(rr)* need not fear having Rh-diseased offspring if her husband is likewise Rh-negative. If the husband is heterozygous *(Rr),* half

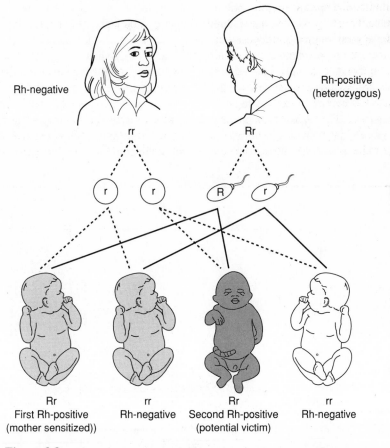

Rh-negative

Rh-positive
(heterozygous)

rr

Rr

r   r   R   r

Rr
First Rh-positive
(mother sensitized))

rr
Rh-negative

Rr
Second Rh-positive
(potential victim)

rr
Rh-negative

Figure 8.8 **Offspring that may arise** from a marriage of an Rh-negative woman and an Rh-positive man (in this case, heterozygous). The mother must be sensitized to the Rh antigen before delivering a child affected with Rh disease.

of the offspring will be Rh-negative *(rr)* and none of these will be afflicted. The other half will be Rh-positive *(Rr),* just like the father, and are potential victims of hemolytic disease. If the Rh-positive father is homozygous *(RR),* then all the children will be Rh-positive *(Rr)* and potential victims. In essence, an Rh-positive child carried by an Rh-negative mother is the setting for possible, though not inevitable, trouble.

The chain of events leading to hemolytic disease begins with the inheritance by the fetus of the dominant *R* allele of the father. The Rh antigens are produced in the red blood cells of the fetus. The fetal red cells bearing the Rh antigens escape through the placental barrier into the mother's circulation, and stimulate the production of antibodies (anti-R) against the Rh antigens on the fetal red cells (fig. 8.9). The mother having produced antibodies, is said to be immunized (or sensitized) against her baby's blood cells. The maternal antibodies hardly ever attain a sufficient concentration during the first pregnancy to harm the fetus. In fact, although fetal cells cross the placenta throughout pregnancy, they enter the maternal circulation in much larger numbers during delivery, when the placental vessels rupture. It is now generally conceded that sensitization of the mother takes place shortly *after* the delivery of the first Rh-positive child. Accordingly, the firstborn is rarely affected, unless the mother had previously developed antibodies from having been transfused with Rh-positive blood or has had a prior pregnancy terminating in an abortion.

The antibodies remain in the mother's system, and may linger for many months or years (fig.8.9). If the second baby is also Rh-positive, the mother may send sufficient antibodies into the fetus' bloodstream to destroy the fetal Rh-positive red cells. The majority of affected fetuses survive for the usual gestation period but are born in critical condition from anemia. Severely anemic individuals are likely to be jaundiced and develop heart failure. A grave threat to the newborn infants is bilirubin, which is a product of red cell destruction. During pregnancy, fetal bilirubin is transported across the placenta and eliminated by the mother. From the time of birth, however, bilirubin accumulates as the affected infant fails to dispose of it. Bilirubin has been shown to be highly toxic to the soft brain tissues; the brain may be permanently damaged.

○ Rh Antigen

⋈ Rh Antibody

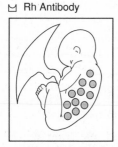 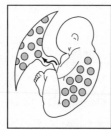 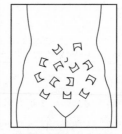 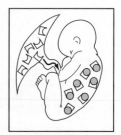

| During pregnancy | At delivery | Months and years later | Subsequent pregnancy |

Figure 8.9 Rh disease in the newborn. The fetus inherits the Rh-positive gene from the father and produces Rh-positive red blood cells. Fetal Rh-positive blood cells enter the Rh-negative mother's bloodstream through placental hemorrhage at delivery, sensitizing her to the Rh-positive antigen and causing her to produce anti-Rh antibodies. The anti-Rh antibodies remain in the mother's bloodstream for many months or years. In a subsequent pregnancy, anti-Rh antibodies in the mother's circulation cross the placenta to react with and destroy fetal red blood cells containing the Rh antigen (if the second fetus is Rh-positive). The fetus is afflicted with Rh disease.

Another aspect of the hemolytic condition, responsible for its original name *(erythroblastosis fetalis),* is the presence of an extraordinarily large number of immature red cells (the erythroblasts) in the circulating blood. It is as if the liver and the spleen, in an attempt to combat the severe anemic condition, produce vast numbers of "unfinished" red blood cells. Unattended, erythroblastosis fetalis leads to stillborn or neonatal death. Many of the erythroblastotic babies are saved by exchange transfusion of Rh-negative blood; others, however, die despite treatment. Exchange transfusions are essentially a flushing-out process, whereby the infant's blood is gradually diluted with Rh-negative donated blood until, at the end of the procedure, most of the infant's circulating blood is problem free. In severe cases, where it has been predicted that the fetus would die before it was mature enough for premature delivery, intrauterine transfusions have been used successfully.

As might have been anticipated, the Rh gene has turned out to be more complex than initially envisioned. There are several variant alleles, and there is a corresponding diversity of antigenic constitutions. This diversity need not concern us here. The most common antigen is the one that was first recognized, known more specifically now as Rh_0, or D. It is the presence of Rh_0 that is tested in ordinary clinical work.

Among Caucasians in the United States, the incidence of Rh-negative persons is approximately 16 percent. In certain European groups, such as the Basques in Spain, the frequency of Rh-negative individuals rises as high as 34 percent. Non-Caucasian populations are relatively free of Rh hemolytic disease. The incidence of Rh-negative persons among the full-blooded American Indians, Eskimos, African blacks, Japanese, and Chinese is 1 percent or less. In contrast, the frequency of Rh-negative African Americans is high (9 percent), which reflects the historical consequences of intermarriages.

We can now appreciate the reasons for viewing Rh disease as an instance of selection *against* the heterozygote. The erythroblastotic infant is always the heterozygote *(Rr).* Each death of an erythroblastotic infant *(Rr)* results in the elimination of one *R* and one *r* gene. In such a situation, where selection continually operates against the heterozygote, the rarer of the two genes should ultimately become lost (or decline to a low level to be maintained solely by mutation). In populations where the *R* gene is much more common that the *r* allele, we should be witnessing a gradual dwindling of the *r* gene.

No decline in the frequency of the *r* gene is evident, however. One counterbalancing factor is the tendency of parents who have lost infants from erythroblastosis to compensate for their losses by having relatively large numbers of children. Thus, if a father is heterozygous *(Rr)* and the mother is homozygous *(rr),* there is an even chance that the infant will be *rr* and unaffected. Each unaffected child born restores two *r* genes lost by the death of two *Rr* erythroblastotic sibs. Accordingly, an excess of homozygous children *(rr)* counterbalances the *r* genes lost through erythroblastosis. This consideration alone overrides the selective force against the heterozygote.

THE CONTROL OF RH DISEASE

In the 1960s Drs. Vincent Freda and John Gorman at the Columbia-Presbyterian Medical Center in New York and Dr. William Pollack at the Ortho Research Foundation in New Jersey sought the means of suppressing the production of antibodies in Rh-negative mothers who had recently delivered an Rh-positive infant. Experiments performed 50 years earlier by the distinguished American bacteriologist, Dr. Theobald Smith, furnished an important clue to the solution. In 1909, Smith arrived at the general principle that passive immunity can prevent active immunity. That is, an antibody given passively by injection can inhibit the recipient from producing its own active antibody. After five years of experimentation and testing, Drs. Freda, Gorman, and Pollack successfully developed an immunosuppressant consisting of a blood fraction (gamma globulin) rich in Rh antibodies. Injected into the bloodstream of the Rh-negative mother no later than three days after the birth of her first Rh-positive child, the globulin-Rh antibody preparation (known as RhoGAM) suppresses the mother's antibody-making activity. Countless numbers of mothers have received the preparation; none have formed active antibodies. More impressive, many of them have delivered a second Rh-positive baby and none of the babies has been afflicted with Rh disease. The evidence is overwhelming that the immunosuppressant is very effective.

The immunosuppressant does not prevent hemolytic disease if maternal antibody is already present by earlier pregnancies or by previous transfusions of Rh-positive blood. For this reason the preparation is administered only to Rh-negative women who do not have anti-Rh in their sera at the time of delivery. With each delivery, opportunity for exposure to Rh-positive fetal cells is repeated. The protection given at the delivery of the first infant does not protect the mother from exposure to antigen received at a subsequent pregnancy. Hence, the immunosuppressant must be given immediately following each pregnancy.

Since the management and prevention of Rh disease have advanced considerably, we have witnessed the end of Rh disease as a troublesome clinical problem. We can also anticipate that the recessive *r* allele will not disappear from the human gene pool, but rather endure. In other words, the *r* allele will be perpetuated by the now-surviving heterozygotes.

Implications of Balanced Polymorphism

According to the classical concept of selection (discussed in chapter 6), a deleterious gene has a harmful effect when homozygous and virtually no expression in the heterozygous state. Deleterious genes will, by selection, be reduced in frequency to very low levels and will be maintained in the population by recurrent mutations. The fittest individuals are homozygous for the normal, or "wild type", allele at most loci. Stated another way, the fittest individuals carry relatively few deleterious genes in the heterozygous state.

In this chapter, we have considered examples of genes that impair fitness when homozygous and actually improve fitness in the heterozygous state. To what extent are individuals heterozygous at their gene loci? Geneticists estimate that 30 percent of the loci in an individual exist in two or more allelic forms. If the alternative alleles at these polymorphic loci are maintained by selection that favors the heterozygote, then low fitness would have to be assigned to an unusually large number of homozygous loci. In other words, an unreasonably high level of unfitness would prevail in the population because of the selective disadvantage of many alleles in the homozygous state.

The possibility exists that most of the alternative alleles at polymorphic loci are *selectively neutral*—that is, the different alleles at one locus confer neither selective advantage nor disadvantage on the individual. Much of the observed variation, then, would represent merely the accumulation of neutral mutations. The notion of selectively neutral alleles has been much debated. Are there mutant alleles whose effects on fitness are not at all different from the more frequent alleles that lead to a normal phenotype? And, can some of these neutral mutant alleles, purely by chance, reach frequencies as high as the frequencies that characterize the state of balanced polymorphism? We will attempt to resolve these vexing queries in subsequent chapters.

GENETIC DRIFT AND GENE FLOW

T he peculiar multilegged condition of the bullfrog (see fig. 1.1) is a rarity in nature. Yet, we witnessed in a particular locality an exceptionally high incidence of this trait. We surmised that the deformity may have been caused by a recessive mutant gene. Since harmful recessive genes in a population tend to be carried mostly in the heterozygous state, the multilegged frogs probably arose from matings of heterozygous carriers. The probability that two or more carriers will actually meet is obviously greater in a small population than in a large breeding assemblage. In fact, the number of matings of carriers of a particular recessive allele in a population is mainly a function of the size of the population. Most populations are not infinitely large, and many fluctuate in size from time to time.

During a period when a population is small, chance matings and segregations could lead to an uncommonly high frequency of a given recessive gene. For example, it is not unthinkable that an unduly harsh winter sharply reduced the size of our particular bullfrog population. By sheer chance, an unusually large proportion of heterozygous carriers of the multilegged condition might

have survived the winter's severity and prevailed as parents in the ensuing spring's breeding aggregation. In this manner, the "multilegged" gene, although not at all advantageous, would occur with an extraordinarily high incidence in the new generation of offspring. Such a fortuitous change in the genetic makeup of a population that may arise when the population becomes restricted in size is known as *genetic drift*.

ROLE OF GENETIC DRIFT

Examination of a natural situation that may be illustrative of genetic drift will lead us into a simplified mathematical consideration of the concept. Coleman Goin, a naturalist at the University of Florida, studied the distribution of pigment variants of a terrestrial frog, known impressively as *Eleutherodactylus ricordi planirostris*. We shall refer to the frog simply by its vernacular name, the greenhouse frog. This frog may possess either of two pigmentary patterns, mottled or striped (fig. 9.1). A unique feature of the greenhouse frog

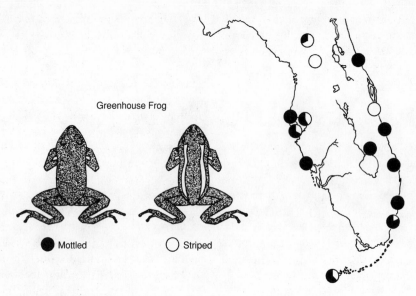

Figure 9.1 Distribtion of the greenhouse frog in Florida, and the relative frequencies of the two pattern variants, *mottled* and *striped*. The populations are small and isolated and differ appreciably in the incidences of mottled and striped forms. The varied frequencies may not be due to natural selection but may represent the outcome of chance fluctuations of genes, or genetic drift. (Based on studies by Coleman Goin.)

is the terrestrial development of its eggs. That is, the eggs need not be submerged completely in water, but can develop in moist earth. This important quality has a considerable bearing on the dispersal of the frogs. Goin reared eggs successfully in a flowerpot two-thirds filled with beach sand and placed in a finger bowl of water. The analysis of a large number of progeny hatched from many different clutches of eggs revealed that the striped pattern is dominant to the mottled pattern.

The greenhouse frog is widespread in Cuba and the Bahama Islands and has only recently become established in Florida. Cuba apparently has been the center of dispersal from which the Florida populations have been derived. The present distribution of the greenhouse frog in Florida consists of a series of small, isolated colonies. As shown in figure 9.1, the proportions of the two patterns vary in different colonies in Florida. In several colonies, only the mottled type occurs. What

can account for the local preponderance of one or the other pattern or even the absence of one of the contrasting patterns?

Goin conjectures that the greenhouse frog was introduced into Florida by means of clutches of eggs accidentally included in shipments of plants from Cuba, a distinct possibility in view of the terrestrial development of the eggs. Thus, a single clutch of introduced eggs could initiate a small colony that, in turn, would establish at the outset a given pattern or proportion of patterns. The presence of only mottled forms in a population may be due to the chance circumstance that only mottled eggs were introduced. Or, perhaps both striped and mottled eggs were included in the shipment, but, by sheer accident, one type was lost in succeeding generations.

It should be understood that Goin has not proved that the unusual distribution and frequency of the two pigment patterns are due solely to chance. The demonstration of genetic drift in any natural population is an extremely difficult task. Genetic drift of the variant pigment patterns is, however, a reasonable explanation.

THEORY OF GENETIC DRIFT

The theory of genetic drift was systematically developed in the 1930s by the distinguished American geneticist Sewall Wright. In fact, the phenomenon of drift is frequently called the *Sewall Wright effect*. The Sewall Wright effect refers specifically to the random fluctuations (or drift) of gene frequencies from generation to generation in a population of small size. Because of the limited size of the breeding population, the gene pool of the new generation may not be at all representative of the parental gene pool from which it was drawn.

The essential features of the process of drift may be seen in the following modest mathematical treatment. Let us assume that the numerous isolated colonies in Florida were each settled by only two frogs, a male and a female, both of constitutions *Aa*. Let us further suppose that each mated pair produces only two offspring. The possible genotypes of the progeny, and the chance associations of the genotypes, are shown in table 9.1.

Several meaningful considerations emerge from table 9.1. For example, the chance that the first offspring from a cross of two heterozygous parents will be *AA* is 1/4. The second event is independent of the first; hence, the chance that the second offspring will be *AA* is also 1/4. The chance that *both* offspring will be *AA* is the product of the separate probabilities of the two independent events, 1/4 × 1/4, or 1/16.

We may now ask: What is the probability of producing two offspring, one *AA* and the other *Aa*, *in no particular order?* From table 9.1, we can see that the chance of obtaining an *AA* individual followed by an *Aa* individual is 2/16. Now, the wording of our question requires that we consider a second possibility, that of an *Aa* offspring followed by an *AA* offspring (also 2/16). These two probabilities must be added together to arrive at the chance of producing the two genotypes irrespective of the order of birth. Hence, in the case in question, the chance is 2/16 + 2/16, or 4/16. In like manner, it may be ascertained that the expectation of obtaining one *AA* and one *aa* offspring (in no given order) is 2/16 and that of producing one *Aa* and one *aa* (in any sequence) is 4/16.

The essential point is that any one of the above circumstances may occur in a given colony. We may concentrate on one situation. The probability that a colony will have only two *AA* offspring is 1 in 16. Thus, by the simple play of chance, the parents initiating the colony might not leave an *aa* offspring. The *a* gene would be immediately lost in

TABLE 9.1	Chance Distribution of Offspring of Two Heterozygous Parents (Aa × Aa)

Genotype of first offspring	Probability of first event	Genotype of second offspring	Probability of second event	Total Probability
AA	1/4	*AA*	1/4	Both offspring *AA*, 1/16
AA	1/4	*Aa*	2/4	*AA* followed by *Aa*, 2/16
AA	1/4	*aa*	1/4	*AA* followed by *aa*, 1/16
Aa	2/4	*AA*	1/4	*Aa* followed by *AA*, 2/16
Aa	2/4	*Aa*	2/4	Both offspring *Aa*, 4/16
Aa	2/4	*aa*	1/4	*Aa* followed by *aa*, 2/16
aa	1/4	*AA*	1/4	*aa* followed by *AA*, 1/16
aa	1/4	*Aa*	2/4	*aa* followed by *Aa*, 2/16
aa	1/4	*aa*	1/4	Both offspring *aa*, 1/16

the population. Subsequent generations descended from the first-generation *AA* individuals would contain, barring mutation, only *AA* types. Chance alone can thus lead to an irreversible situation. A gene once lost could not readily establish itself again in the population. The decisive factor is the size of the population. When populations are small, striking changes can occur from one generation to the next. Some genes may be lost or reduced in frequency by sheer chance; others may be accidentally increased in frequency. Thus, the genetic architecture of a small population may change irrespective of the selective advantage or disadvantage of a trait. Indeed, a beneficial gene may be lost in a small population before natural selection has had the opportunity to act on it favorably.

Founder Effect

When a few individuals or a small group migrate from a main population, only a limited portion of the parental gene pool is carried away. In the small migrant group, some genes may be absent or occur in such low frequency that they may be easily lost. The unique frequencies of genes that arise in populations derived from small bands of colonizers, or "founders," has been called the *founder effect*. This expression emphasizes the conditions or circumstances that foster the operation of genetic drift.

The American Indians afford a possible example of the loss of genes by the founder principle. North American Indian tribes, for the most part, surprisingly lack the allele (designated I^B) that governs type B blood. However, in Asia, the ancestral home of the American Indian, the I^B allele is widespread. The ancestral population of Mongoloids that migrated across the Bering Strait to North America might well have been very small. Accordingly, the possibility exists that none of the prehistoric immigrants happened to be of blood group B. It is also conceivable that a few individuals of the migrant band did carry the I^B allele but they failed to leave descendants.

The interpretation based on genetic drift should not be considered as definitive. The operation of natural selection cannot be flatly dismissed. Most of the North American Indians possess only blood group O, or stated another way, contain only the recessive blood allele *i*. With few exceptions, the North American Indian tribes have lost not only blood group allele I^B but also the allele that controls type A blood (I^A). The loss of both alleles, I^A and I^B, by sheer chance, perhaps defies credibility. Indeed, many modern students of evolution are convinced that some strong selective force led to the rapid elimination of the I^A and I^B genes in the American Indian populations. If this is true, it would offer an impressive example of the action of natural selection in modifying the frequencies of genes in a population.

North American Indians are also known to have a high frequency of albinism. The incidence of albinism among the Cuna Indians of the San Blas Province in Panama is about 1 in 200, which contrasts sharply with the 1 in 20,000 figure for European Caucasians. The Hopi Indians of Arizona and the Zuñi Indians of New Mexico, like the Cuna Indians, are also remarkable in their high numbers of albino individuals.

Since these American Indian populations are small, one might suspect the operation of genetic drift, notwithstanding the adverse effects of the albino allele. It is difficult to imagine, however, that by chance alone this detrimental allele could reach a high frequency independently in several American Indian populations. Charles M. Woolf, geneticist at Arizona State University, has suggested that the high incidence of albinism among the Hopi Indians of Arizona reflects an inimitable form of selection that he terms *cultural selection*. Albinos have been highly regarded in the traditional Hopi society and actually have enjoyed appreciable success in sexual activity. The albino male has been admired, and some have become legendary for leaving large numbers of offspring. Woolf further notes that the fading of old customs among the Hopi Indians is beginning to nullify any reproductive advantage held by albino males in past generations. The frequency of albinism may be expected to decline with the dissolution of the traditional Hopi way of life.

RELIGIOUS ISOLATES

The most likely situation to witness genetic drift is one in which the population is virtually a small, self-contained breeding unit, or *isolate*, in the midst of a larger population. This typifies the Dunkers, a very small religious sect in eastern Pennsylvania. The Dunkers are descendants of the Old German Baptist Brethren, who came to the United States in the early eighteenth century. Bentley Glass, a professor at Johns Hopkins University, studied the community of Dunkers in Franklin County, Pennsylvania, which numbers about 300 individuals. In each generation, the number of parents has remained stable at about 90. The Dunkers live on farms intermingled with the general population, but are genetically isolated by rigid marriage customs. The choice of mates is restricted to members within the religious group.

Glass, with his colleagues, compared the frequencies of certain traits for the Dunker community, the surrounding heterogeneous American population, and the population in western Germany from which the Dunker sect had emigrated two centuries ago. Such a comparison of a small isolate with its large host and parent populations should reveal the effectiveness, if any, of genetic drift. In other words, if the small isolated population shows aberrant gene frequencies as compared to the large parent population, and if the other forces of evolution can be excluded, then the genetic differences can be ascribed to drift.

Analyses were made of the patterns of inheritance of three blood group systems—the ABO blood groups, the MN blood types, and the Rh blood types. In addition, data were accumulated on the incidences of four external traits—namely, the configuration of the ear lobes (which either may be attached to the side of the head or hang free), right- or left-handedness, the presence or absence of hair

on the middle segments of the fingers (mid-digital hair), and "hitch-hiker's thumb," technically termed distal hyperextensibility (fig. 9.2).

The frequencies of many of these traits are strikingly different in the Dunker community from those of the general United States and West German populations. Blood group A is much more frequent among the Dunkers; the O group is somewhat rarer in the Dunkers; and the frequencies of groups B and AB have dropped to exceptionally low levels in the Dunker community. In fact, the I^B allele has almost been lost in the isolate. Most of the carriers of the I^B allele were not born in the community but were converts who entered the isolate by marriage.

A noticeable change has also occurred in the incidences of the M and N blood types in the Dunker community. Type M has increased in frequency, and type N has dwindled in frequency as compared with the incidences of these blood types in either the general United States population or the West German population. Only in the Rh blood groups do the Dunkers conform closely to their surrounding large population.

In physical traits, equally striking differences were found. Briefly, the frequencies of mid-digital hair patterns, distal hyperextensibility of the thumb, and attached ear lobes are significantly lower in the Dunker isolate than in the surrounding American populations. The Dunkers do, however, agree well with other large populations in the incidence of left-handedness. It would thus appear that the peculiar constellation of gene frequencies in the Dunker community—some uncommonly high, others uniquely low, and still others, unchanged from the general large population—can be best attributed to chance fluctuations, or genetic drift.

There is no concurrence of opinion among evolutionists concerning the operation of genetic drift in natural populations, but few would deny that small religious isolates have felt the effect of random sampling. It should be clear, however, that genetic drift becomes ineffectual when a small community increases in size. Fluctuations or shifts

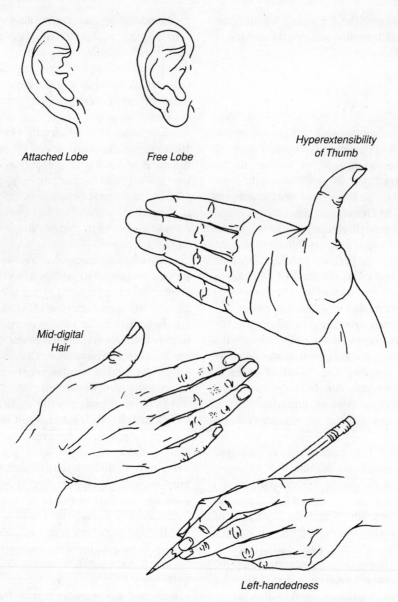

Attached Lobe   Free Lobe

Hyperextensibility of Thumb

Mid-digital Hair

Left-handedness

Figure 9.2 Inheritable physical traits— nature of ear lobes, "hitchhiker's thumb," mid-digital hair, and left-handedness—studied by Bentley Glass and his co-workers in members of the small religious community of Dunkers in Pennsylvania. The distinctive frequency of most of these traits in the Dunker population suggests the operation of genetic drift.

in gene frequencies in large populations are determined almost exclusively by selection.

AMISH OF PENNSYLVANIA

We have seen that gene frequencies in small religious isolates may differ significantly from the original large populations from which the isolates were derived. Another feature of small isolates is the occurrence of rare recessive traits in greater numbers than would be expected from random mating in a large population. This is witnessed among the Old Order Amish societies in the eastern United States.

The Amish sect is an offshoot of the Mennonite Church; both religious groups settled in the United States to escape persecution in Europe in past centuries. The Amish are old fashioned, rural-living people who cultivate the religious life apart from the world. Present-day communities were founded by waves of Amish immigration that began about 1720 and continued until about 1850. The vast majority of Amish live in relatively isolated colonies in Pennsylvania, Ohio, and Indiana. Each community is descended from a small immigrant stock, as attested by the relatively few family names in a given community. Analyses by the geneticist Victor A. McKusick of Johns Hopkins University have shown that eight names account for 80% of the Amish families in Lancaster County, Pennsylvania. Other Amish communities also are characterized by a high frequency of certain family names, as table 9.2 shows.

Marriages have been largely confined within members of the Amish sect, with a resulting high degree of consanguinity. Marriages of close relatives have tended to promote the meeting of two normal, but carrier, parents. Four recessive disorders manifest themselves with uncommonly high frequencies, each in a different Amish group: the Ellis-van Creveld syndrome, pyruvate kinase-deficient hemolytic anemia, Hemophilia B (Christmas disease), and a form of limb-girdle muscular dystrophy (Troyer syndrome).

We may consider in some detail the Ellis-van Creveld syndrome, which occurs in the Lancaster County population (see fig.1.6). Fifty-two affected persons have been identified in 30 sibships, most of which have unaffected parents. Pedigree analysis has revealed that Samuel King and his wife, who immigrated in 1744, are ancestral to all parents of the sibships. Either Samuel King or his wife carried

TABLE 9.2	Old Order Amish Family Names in Three American Communities*				
Lancaster Co., Pa.		**Holmes Co., Ohio**		**Mifflin Co., Pa.**	
Stolzfus**	23%	Miller	26%	Yoder	28%
King	12%	Yoder	17%	Peachey	19%
Fischer	12%	Troyer	11%	Hostetler	13%
Beiler	12%	Hershberger	5%	Byler	6%
Lapp	7%	Raber	5%	Zook	6%
Zook	6%	Schlabach	5%	Speicher	5%
Esh***	6%	Weaver	4%	Kanagy	4%
Glick	3%	Mast	4%	Swarey	4%
	81%		77%		85%
Totals:					
1,106 families, 1957		1,611 families, 1960		238 families, 1951	

*From data compiled by Victor A. McKusick of John Hopkins University.

**Including Stolzfoos.

***Including Esch.

the recessive gene. None of their children were affected, but subsequent generations were. Evidently, previously concealed detrimental recessive genes are brought to light by the increased chances of two heterozygotes meeting in a small population.

CONSANGUINITY AND GENETIC DRIFT

Small populations not only provide opportunities for genetic drift but for consanguinity as well. As witnessed among the Amish, offspring afflicted with recessive disorders arise more often from unions of close relatives than from marriages of unrelated persons. Yet the Amish sternly frown on first-cousin marriages. Nonetheless, although consanguineous marriages are not made by choice, the limited size of the Amish population restricts the availability of potential mates and virtually forces marriages of close relatives. Indeed, second-cousin and third-cousin marriages are rather common among the Amish. The relationship between genetic drift and inbreeding thus becomes clear: the closed nature of a small population creates a situation wherein few nonrelatives are present in the population. Because matings of genetically related individuals increase the probability of homozygosity in the progeny, a relatively large number of rare recessive homozygotes are expected, and do occur.

The smaller the population and the longer it has been isolated, the greater the chance that most members of the population are related to each other through common ancestors. This is exemplified by the isolated island of Tristan da Cunha in the South Atlantic. This small island is of historical interest. In 1816, a British military garrison was established on Tristan for the sole purpose of safeguarding against the escape of Napoleon from the neighboring, though distant, St. Helena. Within a few months, it became glaringly apparent that Tristan was inconsequential to Napoleon's safekeeping. The battery of soldiers was hastily withdrawn, but a Scots corporal, William Glass, his family, and a few soldiers remained behind. From 1817 until 1908, women from St. Helena

and shipwrecked sailors joined the community to bring the total to a mere 15 individuals. The present population of nearly 270 traces their origin to the 15 early settlers.

The English geneticist D. F. Roberts has calculated that in Tristan today the probability is high that any two young individuals contemplating marriage share numerous genes due to the common ancestry of their parents. Roberts stresses that the relatively high level of consanguinity on the island does not reflect a conscious preference for marriage between close relatives. Rather, random mating prevails, but the marriages prove to be consanguineous because most of the potential mates are already relatives by common descent.

GENE FLOW

A rich archeological record reveals appreciable movement on the part of early human populations. Some migrations were sporadic, in small groups; others were more or less continual streams involving large numbers of people. Large-scale immigrations followed by interbreeding have the effect of introducing new genes to the host populations. The diffusion of genes into populations through migrations is referred to as *gene flow*.

The graded distribution of the I^B blood-group allele in Europe represents the historical consequence of invasions by Mongolians who pushed westward repeatedly between the sixth and sixteenth centuries (fig. 9.3). There is a high frequency of the I^B allele in central Asia. In Europe, the frequency of the I^B gene diminishes steadily from the borders of Asia to a low level of 5 percent or less in parts of Holland, France, Spain, and Portugal. The Basque peoples, who inhabit the region of the Pyrenees in Spain and France, have the lowest frequency of the I^B allele in Europe—below 3 percent. From a biological standpoint, the Basque community of long standing is a cohesive, endogamous mating unit. The exceptionally low incidence of the I^B gene among the Basques may be taken to indicate that there has been little intermarriage with surrounding populations. It is possible that a few centuries ago

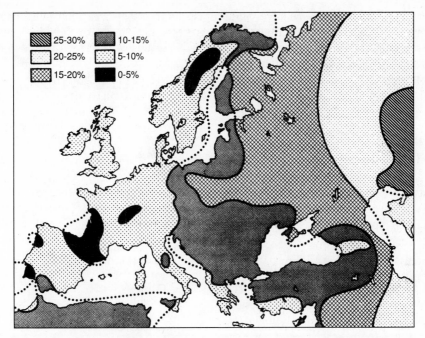

Figure 9.3 Gradient of frequencies of the I^B blood-group gene from central Asia to western Europe. (Based on studies by A. E. Mourant.)

the I^B allele was completely absent from the self-contained Basque community.

The exchange of genes between populations may have dramatic consequences. Until recently, Rh disease was virtually unknown in China. Less than 70 years ago, all Chinese women were Rh-positive *(RR)*. However, intermarriage between immigrant Americans and the native Chinese has led to the introduction of the Rh-negative allele *(r)* in the Chinese population. No Rh disease would be witnessed in the immediate offspring of American men and Chinese women. By contrast, all marriages of Rh-negative American women *(rr)* and Rh-positive Chinese men *(RR)* would be of the incompatible type. All children by these Chinese fathers would be Rh-positive *(Rr)* and potential victims of hemolytic disease.

Whereas American immigrants introduce the Rh-negative allele *(r)* into Chinese populations where it formerly was not present, Chinese immigrants (all of whom are *RR*) introduce more Rh-positive genes *(R)* into the American populations, thus diluting the Rh-negative gene pool in the United States. Initially, the Rh-positive Chinese men *(RR)* married to Rh-negative American women *(rr)* would result in an increased incidence of Rh-diseased infants. In later generations, however, the frequency of Rh-negative women in the United States would be lower, inasmuch as women of mixed Chinese-American origin would be either *RR* or *Rr*, predominantly the former. Thus, in the United States, the long-range effect of Chinese-American intermarriage is a reduction in the incidence of hemolytic disease of the newborn.

CAUCASIAN GENES IN AFRICAN AMERICANS

In the United States, from a beginning more than 250 years ago, there has occurred an admixture of American Caucasians with the African Americans who were brought from Africa to the New World as slaves. The overall Caucasian contribution to the African American gene pool has been estimated at approximately 20 percent. This percentage must be considered as a gross estimate, inasmuch as different African American populations in the United States have undergone varied degrees of admixture.

The geneticist T. Edward Reed of the University of Minnesota established the frequency with which a particular blood-group gene appears in American Caucasians, African Americans, and natives of West Africa. The specific gene is the Fy^a allele of the Duffy blood group system, which may be referred to simply as "the Duffy factor." As seen in table 9.3, the Duffy factor is virtually absent in the West African populations from which most of the original immigrant slaves were derived. The highest frequency of the Duffy factor in the stem populations in Africa is an exceedingly low 0.04, whereas the value for representative Caucasian populations in the United States is 0.4, or ten times as great (table 9.3). Among African Americans residing in New York City, Detroit, and Oakland, the gene reaches a frequency of about 0.10. This signifies that the magnitude of Caucasian ancestry in African Americans in these localities may be as high as 26 percent. In contrast, the frequency of the Duffy factor among African Americans in Charleston, South Carolina, is 0.02, which indicates that the African American population in Charleston has only a small amount of Caucasian ancestry. Thus, although the extent of Caucasian-African American hybridization has been variable, the calculations indicate that various African American

TABLE 9.3	**Duffy Blood Factor (Fy^a Gene) in West African, American Caucasian , and African American Populations**

Population	Fequency of Duffy Factor	Percentage of Caucasian Contribution
A. West African		
Liberia	0.00	—
Ivory Coast	0.04	—
Upper Volta	0.00	—
Dahomey	0.00	—
Ghana (Accra)	0.00	—
Nigeria (Lagos)	0.00	—
B. American Caucasian		
Oakland, California	0.43	
Evans and Bullock		
Counties, Georgia	0.42	
C. African American		
1. Non-Southern		
New York, New York	0.08	18.9
Detroit, Michigan	0.11	26.0
Oakland, California	0.09	22.0
2. Southern		
Charleston,		
South Carolina	0.02	3.7
Evans and Bullock		
Counties, Georgia	0.05	10.6

Based on data compiled by T. Edward Reed of the University of Minnesota.

populations derive between 4 percent and 26 percent of their genes from Caucasian ancestors, and that these Caucasian genes have been introduced through hybridization since 1700.

CHAPTER

10

RACES AND SPECIES

Any large assemblage of a particular organism is generally not distributed equally nor uniformly throughout its territory or range in nature. A widespread group of plants or animals is typically subdivided into numerous local populations, each physically separated from the others to some extent. The environmental conditions in different parts of the range of an organism are not likely to be identical. We may thus expect that a given local population will consist of genetic types adapted to a specific set of prevailing environmental conditions. The degree to which each population maintains its genetic distinctness is governed by the extent to which *interbreeding* between the populations occurs. A free interchange of genes between populations tends to blur the differences between the populations. But what are the consequences when gene exchange between populations is greatly restricted or prevented? This chapter addresses itself to this question.

VARIATION BETWEEN POPULATIONS

Our first consideration is to demonstrate that heritable variations exist among the various breeding populations in different geographical localities of an organism. Jens Clausen, David Keck, and William Hiesey of the Carnegie Institution of Washington at Stanford, California, demonstrated that each of the populations of the yarrow plant *Achillea lanulosa* from different parts of California are each adapted to their respective habitat. As figure 10.1 shows, the variations in height of the plant are correlated with altitudinal differences. The shortest plants are from the highest altitudes, and the plants increase in height in a gradient fashion with decreasing altitude. The term *cline*, or character gradient, has been applied to such situations where a character varies more or less continuously with a gradual change in the environmental terrain.

The observation by itself that the yarrow plants are phenotypically dissimilar at different elevations does not indicate that they are genetically different. If the observed variations are claimed as local adaptations resulting from natural selection, then a hereditary basis for the differences in height should be demonstrated. It is often difficult to obtain data that disclose the hereditary nature of population differences. In this respect, the studies of Clausen and his co-workers are

109

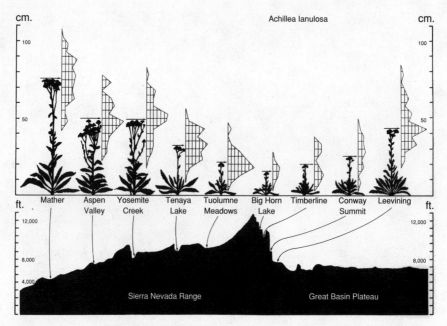

Figure 10.1 **Clinal variation** in the yarrow plant, *Achillea lanulosa*. The increase in height of the plant is more or less continuous with decreasing altitude. The plants shown here are representatives from different populations in the Sierra Nevada Mountains of California that were grown in a uniform garden at Stanford, California. Each plant illustrated is one of the average height for the given population; the graph adjacent to the plant reveals the distribution of heights within the population.

(From Clausen, Keck, and Hiesey, *Carnegie Institution of Washington Publication 581,* 1948.)

commendable. The plants shown in figure 10.1 had actually been grown together in a uniform experimental garden at Stanford, California. Transplanted from various localities, the plants developed differently from one another in the same experimental garden, revealing that each population had evolved its own distinctive complex of genes.

RACES

The variation pattern in organisms may be discrete, or discontinuous, particularly when the populations are separated from each other by pronounced physical barriers. This is exemplified by the varieties of the carpenter bee *(Xylocopa nobilis)* in the Celebes and neighboring islands of Indonesia

(fig. 10.2). As shown by the studies of J. van der Vecht of the Museum of Natural History at Leiden in the Netherlands, there are three different varieties of carpenter bee on the mainland of Celebes and at least three kinds on the adjacent small islands. These geographical variants differ conspicuously in the coloration of the small, soft hairs that cover the surface of the body. The first abdominal segment is invariably clothed with bright yellow hairs. However, each variety has evolved a unique constellation of color on the other abdominal segments and also on the thorax.

The variations in the carpenter bees within and between islands are well defined and easily distinguishable. One may refer to populations with well-marked discontinuities as *races*. Races are simply geographical aggregates of populations

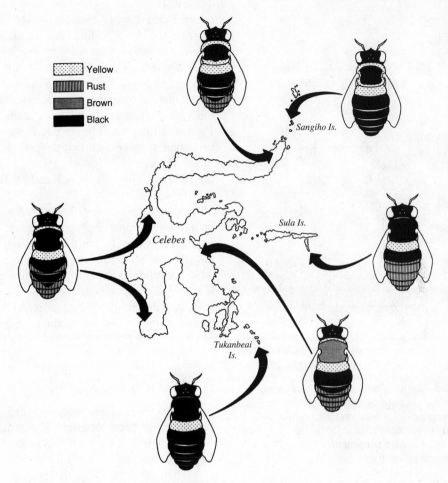

Figure 10.2 Geographic variation of color patterns in females of the carpenter bee, *Xylocopa nobilis*, in the Celebes and neighboring islands in Indonesia. Each geographic race has evolved a distinctive constellation of colors.
(Based on studies by J. van der Vecht.)

that differ in the incidence of genetic traits. How genetically different two assemblages of populations must be to warrant racial designations is an open question.

Some of the problems inherent in delimiting races are exemplified by the different temperature-adapted populations of the North American leopard frog, *Rana pipiens*. John A. Moore, then at Columbia University and later at the University of California, tested the effects of temperature on the

development of the embryos of frogs from widely different localities. He wished to ascertain the limits of temperatures that the embryos can endure or tolerate. The findings on four different geographic populations in the eastern United States are shown in figure 10.3.

Embryos of northern *Rana pipiens* populations are more resistant to low temperatures and less tolerant of high temperatures than are embryos from southern populations. Embryos of

Vermont
5° – 28°

New Jersey
5° – 28°

Northern Florida
9° – 33°

Southern Florida
11° – 33°

Figure 10.3 Limits of temperature tolerance of embryos of the leopard frog *(Rana pipiens)* from different geographical populations. Embryos of northern populations are more resistant to low temperatures and less tolerant of high temperatures than are embryos from southern populations.

(Based on studies by John A. Moore.)

populations from Vermont and New Jersey have comparable ranges of temperature tolerances. These northern embryos can resist temperatures as low as 5°C. Embryos from Florida differ markedly from those of northern populations. Embryos from southern Florida (latitude 27°N) can tolerate temperatures as high as 33°C, but are very susceptible to low temperatures. Hence, northern and southern populations have become adapted to different environments in their respective territories.

We may refer to the northern populations as the cold-adapted race of the leopard frog, and designate the southern populations as the warm-

adapted race. It is evident that we are being arbitrary in drawing a fine line of demarcation between northern and southern races. Data are presently lacking for the geographically intermediate populations, but further studies will probably reveal that the temperature adaptations of the frog embryos change gradually from north to south. Even with the present information, one may wish to recognize more than two temperature-adapted races, or perhaps, as some investigators firmly argue, refrain completely from making racial designations.

Races may be best thought of as units of organization below the species level. In other words, races may be considered as stages in the transformation of populations into species. But what constitutes a species? Up to this point, we have assiduously avoided the use of the term *species*. A discussion of the process leading to the formation of species will facilitate understanding of the term itself.

FORMATION OF SPECIES

Let us imagine a large assemblage of land snails subdivided in three geographical aggregations or races, *A, B,* and *C,* each adapted to local environmental conditions (fig. 10.4). There are initially no gross barriers separating the populations from each other, and where *A* meets *B* and *B* meets *C,* interbreeding occurs. Zones of intermediate individuals are thus established between the races, and the width of these zones depends on the extent to which the respective populations intermingle. It is important to realize that races are fully capable of exchanging genes with one another.

We may now visualize (fig. 10.4) some striking feature, such as a great river, forging its way through the territory and effectively isolating the land snails of race *C* from those of *B.* These two assemblages may be spatially separated from each other for an indefinitely long period of time, affording an opportunity for race *C* to pursue its own independent evolutionary course. Two populations that are geographically separated, like *B*

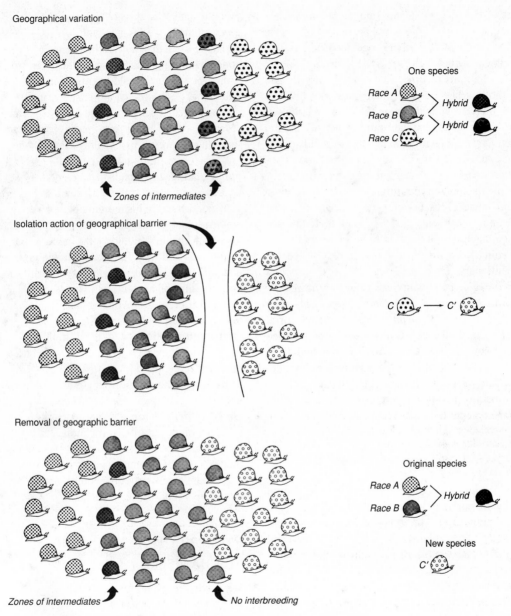

Figure 10.4 Model for the process of geographic speciation. Members of populations (or race) *C* had diverged genetically during geographical isolation in ways that have made them reproductively incompatible with race *B* when they met again. Race *C* has thus transformed into a new species, *C'*.

and *C* in our model, are said to be *allopatric*. (Technically speaking, *A* and *B* are also allopatric since, for the most part, they occupy different geographical areas.)

After eons of time, the river may dry up and the hollow bed may eventually become filled in with land. Now, if the members of populations *B* and *C* were to extend their ranges and meet again, one of two things might happen. The snails of the two populations might freely interbreed and establish once again a zone of intermediate individuals. Or, the two populations might no longer be able to interchange genes because they have become so different genetically that they cannot. If the two assemblages can exist side by side without interbreeding, then the two groups have reached the evolutionary status of separate species. *A species is a breeding community that preserves its genetic identity by its failure to exchange genes with other such breeding communitites.* In our pictorial model (fig. 10.4), race *C* has become transformed into a new species, *C'*. Two species *(A-B and C')* have now arisen where formerly only one existed. It should be noted that races *A* and *B* are treated as members of a single species since no barriers to gene exchange exist between them.

NOMENCLATURE

The scientific names that the taxonomist would apply to our populations of land snails deserve special comment. The technical name of a species consists of two words, in Latin or in latinized form. An acceptable designation of the original species of land snails depicted in figure 10.4 would be *Helix typicus*. The first word is the name of a comprehensive group, the genus, to which land snails belong; the second word is a name unique to the species. The taxonomist would be obliged to create a different latinized second name for the newly derived species of the land snail, the *C'* population in figure 10.4. This new species might well be called *Helix varians*. The name of the genus remains the same since the two species are closely related. The genus, therefore, denotes a

group of interrelated species. Taxonomists choose a given latinized species name for a variety of reasons, and more often than not, the latinized name does not connote much information about the organism itself. Thus, the student should not imagine that the key features of each species are encoded in the name, any more than a person's given name is particularly revealing of his or her attributes. The names are important, however, in revealing relationships.

The binomial ("two-named") system of nomenclature, universally accepted, was devised by the Swedish naturalist Carolus Linnaeus (born Karl von Linné) in his monumental work, *Systema Naturae*, first published in 1735. Convention dictates that the first letter of the generic name be capitalized and that the specific name begin with a small letter. It is also customary to print the scientific name of a species in italics, or in a type that is different from that of the accompanying text. A modern refinement of the Linnaean system is the introduction of a third italicized name, which signifies the subspecies. Geographical races are recognized taxonomically as subspecies. Thus, it would be appropriate to designate races *A* and *B* (fig.10.4) as *Helix typicus elegans* and *Helix typicus eminens,* respectively. Such a species composed of two (or more) subspecies is said to be *polytypic*. A *monotypic* species is one that is not differentiated into two or more geographical races or subspecies. *Helix varians* would be a monotypic species.

REPRODUCTIVE ISOLATING MECHANISMS

We have seen that two populations (or races), while spatially separated from each other, may accumulate sufficient genetic differences in isolation that they would no longer be able to interchange genes if they came into contact with one another. When the geographical barrier persists, it is difficult to judge the extent to which the two allopatric populations have diverged genetically from each other. Only when the two populations come together again does it become apparent whether or not they have changed in ways that would make them reproductively incompatible. Two populations that come to occupy the same territory are called *sympatric*. The agencies that prevent interbreeding between sympatric species are known as *reproductive isolating mechanisms.*

Reproductive isolating mechanisms are of different types, and one or more of the types may be found separating two species. The various types may be grouped into two broad categories. One category includes the prezygotic (or premating) mechanisms, which serve to prevent the formation of hybrid zygotes. The other category encompasses the postzygotic (or postmating) mechanisms, which act to reduce the viability or fertility of hybrid zygotes. The specific types of isolating mechanisms under these two groupings can be listed as follows:

A. Prezygotic (premating) mechanisms
   1. Habitat (ecological) isolation
   2. Seasonal (temporal) isolation
   3. Sexual (ethological) isolation
   4. Mechanical isolation
   5. Gametic isolation

B. Postzygotic mechanisms
   1. Hybrid inviability
   2. Hybrid sterility
   3. Hybrid breakdown

Two related species may live in the same general area but differ in their ecological requirements. The scarlet oak *(Quercus coccinea)* of eastern North America grows in moist or swampy soils, whereas the black oak *(Quercus velutina)* is adapted to drier soils (fig. 10.5). The two kinds of oak are thus effectively separated by different *ecological* or *habitat* preferences. Two sympatric species may also retain their distinctness by breeding at different times of the year *(seasonal isolation)*. Evidently, cross-fertilization is not feasible between two species of frogs that release their gametes on different months even in the same pond, or between two species of pine that shed their pollen in different periods. The breeding seasons of two species may overlap, but interbreeding may not occur because of the lack of mutual attraction between the sexes of the two species *(sexual isolation)*. Among birds, for example, elaborate courtship rituals play important roles in species recognition and the avoidance of interspecific matings.

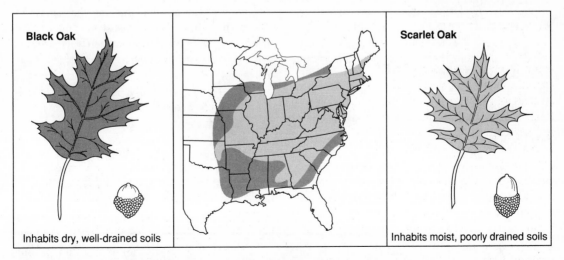

Figure 10.5 **Ecological isolation** is exemplified by the different habitat requirements of two species of oak. Although both species occur in the eastern United States, the scarlet oak *(Quercus coccinea)* is adapted to the moist bottom lands, whereas the black oak *(Quercus velutina)* is adapted to the dry upland soils.

In many insects, interbreeding of species is hindered by differences in the structure of the reproductive apparatus *(mechanical isolation)*. Copulation is not possible because the genitalia of one species is physically incompatible with the genitalia of the other species. In fact, several closely related species of insects can often be accurately classified by their distinctive genitalia. In some instances, the male of one species may inseminate the female of the other species, but the sperm cells may be inviable in the reproductive tract of the female. This form of *gametic isolation* is not unique to animals; in plants, such as the Jimson weed *(Datura)*, the sperm-bearing pollen tube of one species encounters a hostile environment in the flower tissue of the other species and is unable to reach the egg.

Cross-fertilization between two species may be successful, but the hybrid embryos may be abnormal or fail to reach sexual maturity *(hybrid inviability)*. For example, two species of the chicory plant, *Crepis tectorum* and *Crepis capillaris,* can be crossed, but the hybrid seedlings die in early development. Crosses between the bullfrog, *Rana catesbeiana,* and the green frog, *Rana clamitans,* result in inviable embryos. In certain hybrid crosses, such as between females of the toad species *Bufo fowleri* and males of *Bufo valliceps,* the hybrids may survive but are completely sterile *(hybrid sterility)*. The familiar example of hybrid sterility is the mule, the offspring of a male ass and a female horse. In some situations, the F_1 hybrids appear to be vigorous and fertile, but the viability of a subsequent generation is very reduced. Such a case of *hybrid breakdown* has been described in several species of cotton—*Gossypium hirsutum* and *Gossypium barbadense*. These species produce normal, fertile F_1 hybrids, but the majority of the F_2 hybrid cotton seedlings fail to germinate.

In essence, two populations can remain distinct, and be designated as species, when gene exchange between them is prevented or limited by one or more reproductive isolating mechanisms. More often than not, we are unable to obtain direct evidence for the presence or absence of inter-breeding in nature between two groups. The degree of reproductive isolation is then indirectly gauged by the extent to which the members of two populations differ in morphological, physiological, and behavioral characteristics. Two populations that are morphologically very dissimilar are likely to be distinct species. It should be understood, however, that the level of morphological differentiation cannot be used as a reliable criterion of a species. For example, two species of fruit fly, *Drosophila pseudoobscura* and *Drosophila persimilis*, are reproductively isolated but are almost indistinguishable on morphological grounds.

ORIGIN OF ISOLATING MECHANISMS

How do reproductive isolating mechanisms arise? In the 1940s, John A. Moore undertook a series of instructive evolutionary studies on the leopard frog, *Rana pipiens*. The leopard frog is widely distributed in North America, ranging from northern Canada through the United States and Mexico into the lower reaches of Central America. Moore obtained leopard frogs from different geographical populations and crossed them in the laboratory. When frogs from the northeastern United States (Vermont) were crossed with their southerly distributed lowland relatives in eastern Mexico (Axtla in San Luis Potosi), the hybrid embryos failed to develop normally. Thus, of this species, the members geographically most distant from one another have diverged genetically to the extent that when cross-mated in the laboratory they are incapable of producing viable hybrids.

It must be admitted that the possibility of a Vermont frog crossing with a Mexican frog in nature is extremely remote. It took a biologist to bring these two frogs together. Yet, it is just this point that emphasizes that an isolating mechanism, such as hybrid inviability, does not develop for the effect itself; it is simply the natural consequence of sufficient genetic differences having accumulated in two populations during a long geographical separation. The late Hermann J. Muller of Indiana University was among the first to suggest that isolating mechanisms originate as a by-product of genetic divergence of allopatric populations. The genetic changes that arise to adapt one population to particular environmental conditions may also be instrumental in reproductively isolating that population from other populations that are themselves developing adaptive gene complexes. Indeed, the embryos of Vermont leopard frogs differ considerably in their range of temperature tolerance from embryos of eastern Mexican frogs. It might well be, then, that the embryonic defects in hybrids between these northern and southern frogs are associated with the different temperature adaptations of the parental eggs.

If the Vermont and Mexican leopard frogs were ever to meet in nature, any intercross between them would lead to the formation of inviable hybrids. This would represent a wastage of reproductive potential of the parental frogs. Theodosius Dobzhansky advanced the engaging hypothesis that under such conditions, natural selection would promote the establishment of isolating mechanisms that would guard against the production of abnormal hybrids. In frogs, a normal mating or a mismating in a mixed population depends principally on the discrimination of the female. The reproductive potential is obviously lower for an undiscriminating female than for a female who leaves normal offspring. If the tendency to mismate is heritable, then the genes responsible for this tendency will eventually be lost or sharply reduced in frequency by elimination of the indiscriminate females, an elimination effectively accomplished by the inviability of their offspring. Thus, the continual propagation of females that most resist the attentions of "foreign" males will lead eventually to a situation in which mismatings do not occur and abnormal hybrids are no longer produced.

Karl Koopman, an able student of Theodosius Dobzhansky, tested the thesis that natural selection tends to strengthen, or make complete, the reproductive isolation between two species coexisting in the same territory. Koopman used for experimentation two species of fruit fly, *Drosophila pseudoobscura* and *Drosophila persimilis*. In nature, sexual selection between these two sympatric species is strong, and interspecific matings do not occur. However, in a mixed population in the laboratory, particularly at low temperatures, mismatings do take place. Koopman accordingly brought together members of both species in an experimental cage and purposely kept the cage at a low temperature (16°C). Hybrid flies were produced and were viable, but Koopman in effect made them inviable by painstakingly removing them from the breeding cage when each new generation emerged. Over a period of several generations, the production of hybrid flies dwindled markedly and mismatings in the population cage were substantially curtailed. This is a dramatic demonstration of the efficacy of selection in strengthening reproductive isolation between two sympatric species.

HUMANS: A SINGLE VARIABLE SPECIES

There is only one present-day species of human, *Homo sapiens*. Different populations of humans can interbreed successfully and, in fact, do. The extensive commingling of populations renders it difficult, if not impossible, to establish discrete racial categories in humans. Races, as we have seen, are geographically defined aggregates of local populations. The populations of humankind are no longer sharply separated geographically from one another. Multiple migrations of peoples and innumerable intermarriages have tended to blur the genetic contrasts between populations. The boundaries of human races, if they can be delimited at all, are at best fuzzy, ever shifting with time.

The term *race* is regrettably one of the most abused words in the English vocabulary. The biologist views a race as synonymous with a geographical subspecies; a race or subspecies is a genetically distinguishable subgrouping of a species. It is exceedingly important to recognize that a race is *not* a community based on language, literature, religion, nationality, or customs. There are Aryan languages, but there is no Aryan race. Aryans are people of diverse genetic makeups who speak a common language (Indo-European). *Aryan* is therefore nothing more than a linguistic designation. In like manner, there is a Jewish religion, but not a Jewish race. And there is an Italian nation, but not an Italian race. A race is a reproductive community of individuals occupying a definite region, and in one and the same geographical region may be found Aryans, Jews, and Italians. Every human population today consists of a multitude of diverse genotypes. A pure population or race, in which all members are genetically alike, is non-existent.

11

CATACLYSMIC EVOLUTION

The process leading to the formation of a new species generally extends over a great reach of time. As we have seen, the origin of new species involves a long period of geographical isolation and the long-term influences of natural selection. A sudden and rapid emergence of a new kind of organism is scarcely imaginable. Yet, a natural mechanism does exist whereby a new species can arise rather abruptly. The process is associated with the phenomenon of *polyploidy,* or the multiplication of the chromosome complement of an organism. Species formation through polyploidy has occurred almost entirely, if not exclusively, in the plant kingdom. Many of our valuable cultivated crop plants, such as wheat, oats, cotton, tobacco, and sugar cane, trace their origin to this cataclysmic or explosive type of evolution.

WHEAT

The domestic wheats and their wild relatives have an intriguing evolutionary history. There are numerous species of wheat, all of which fall into three major categories on the basis of their chro-

mosome numbers. The most ancient type is the small-grain einkorn wheat, containing 14 chromosomes in its body (somatic) cells. There are two species of einkorn wheat, one wild and the other cultivated, and both of them may be found growing in the hilly regions of southeastern Europe and southwestern Asia. Cultivated einkorn has slightly larger kernels than the wild form, but the yields of each are low and the grain is used principally for feeding cattle and horses.

Another assemblage of wheat, once widely grown, is the emmer series, of which there are at least six species. The chromosome number in the nuclei of somatic cells of emmer wheats is 28. These varieties, found in Europe and the United States, are used today principally as stock feed, although one of them, called durum wheat, is of commercial value in the production of macaroni and spaghetti.

The most recently evolved, and by far the most valuable agriculturally, are the bread wheats. The bread wheats have not been known to occur in the wild state; all are cultivated types. The bread wheats have 42 chromosomes. These wheats, high in protein content, comprise almost 90 percent of all the wheat harvested in the world today.

The various species of wheat thus fall into three major groups, with 14, 28, and 42 chromosomes, respectively. A list of the representatives of these three groups is given in table 11.1.

ORIGIN OF WHEAT SPECIES

Virtually all authorities are agreed on the sequence of evolutionary events depicted in figure 11.1. Einkorn wheat, possessing a chromosome number of 14, was doubtless one of the ancestral parents of the 28-chromosome emmer assemblage. A most remarkable, but generally accepted, thesis is that the other parent was not a wheat at all, but rather *Aegilops speltoides,* a wild grass with 14 chromosomes. This wild grass parent occurs as a common weed in the wheat fields of southwestern Asia. The cross of einkorn wheat and the wild grass would yield an F₁ hybrid that possesses 14 chromosomes, 7 from each parent. We may designate the 7 chromosomes from one parent species as set (or genome) *A,* and the 7 from the other parent species as set (or genome) *B.* Accordingly, the F₁ hybrids would have the *AB* genomes.

If the chromosome complement in the hybrid accidentally doubled, then the hybrid would con-

tain 28 instead of 14 chromosomes and pass on the doubled set of chromosomes to its offspring. Such an event, strange as it may seem, accounts for the emergence of the 28-chromosome emmer wheat. This new species is characterized as having the *AABB* genomes.

In turn, the 28-chromosome emmer wheat was the ancestor of the 42-chromosome bread wheat. In the early 1900s, the British botanist John Percival hazarded the opinion that the bread wheat group arose by hybridization of a species of wheat of the emmer group (28 chromosomes) and goat grass, *Aegilops squarrosa,* a useless weed commonly found growing in the wheat fields in the Mediterranean area. Although this startling suggestion was initially viewed with skepticism, it is currently conceded that Percival was correct. *Aegilops squarrosa* possesses 14 chromosomes, and thus would transmit 7 of its chromosomes (set *C*) to the hybrid. The hybrid would contain 21 chromosomes (sets *ABC*), having received 14 (sets *AB*) from its emmer wheat parent. The subsequent duplication in the hybrid of each chromosome set provided by the parents would result in a 42-chromosome wheat species (*AABBCC*).

The initial F₁ hybrid between einkorn wheat and *Aegilops speltoides* (or between emmer wheat and *Aegilops squarrosa*) is sterile, but when the chromosome complement doubles, then

TABLE 11.1 | **Species of Wheat (*Triticum*)**

14 Chromosomes	28 Chromosomes	42 Chromosomes
T. aegilopoides (Wild Einkorn)	*T. dicoccoides* (Wild Emmer)	*T. aestivum* (Bread Wheat)
T. monococcum (Cultivated Einkorn)	*T. dicoccum* (Cultivated Emmer)	*T. sphaerococcum* (Shot Wheat)
	T. durum (Macaroni Wheat)	*T. compactum* (Club Wheat)
	T. persicum (Persian Wheat)	*T. spelta* (Spelt)
	T. turgidum (Rivet Wheat)	*T. macha* (Macha Wheat)
	T. polonicum (Polish Wheat)	

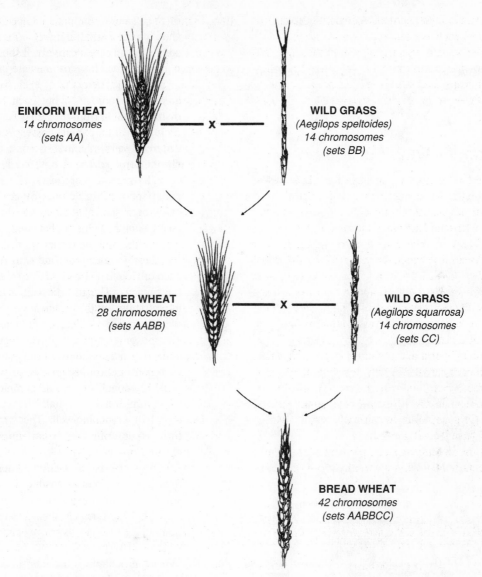

Figure 11.1 Evolution of wheat. Emmer wheat resulted from the hybridization of einkorn wheat with a wild grass, *Aegilops speltoides*. The common bread wheat is the product of hybridization of emmer wheat with a useless and noxious species of wild grass, *Aegilops squarrosa*. The hybridizations were followed by chromosome doubling, a phenomenon discussed in the text.

a fully fertile species arises. Is it to be expected that the F_1 hybrid would be sterile? And what would account for the fertility of the hybrid when chromosome doubling occurred? This requires a deeper look into the phenomenon of polyploidy, to which we shall now turn.

MECHANISM OF SPECIATION BY POLYPLOIDY

Figure 11.2 illustrates the underlying basis of the fertility of a formerly sterile hybrid resulting from the doubling of its chromosome number. For ease

SPECIATION BY POLYPLOIDY

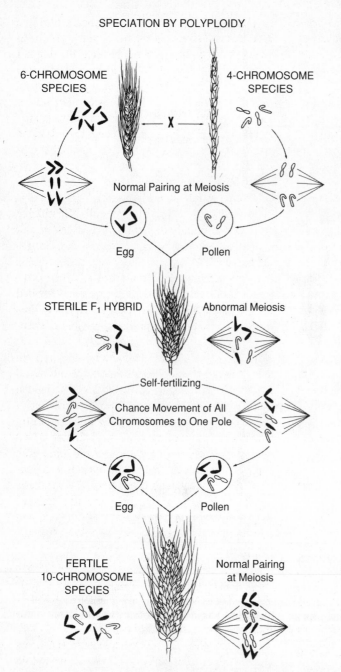

6-CHROMOSOME SPECIES

4-CHROMOSOME SPECIES

Normal Pairing at Meiosis

Egg   Pollen

STERILE F₁ HYBRID   Abnormal Meiosis

Self-fertilizing
Chance Movement of All Chromosomes to One Pole

Egg   Pollen

FERTILE 10-CHROMOSOME SPECIES

Normal Pairing at Meiosis

of presentation, the parental species are shown with a small number of chromosomes, 6 and 4, respectively. It should be noted that the chromosomes are present in pairs. The members of each pair are alike or homologous, but each pair is distinguishable from the other. Thus, the 6-chromosome parent possesses three different pairs of chromosomes; the 4-chromosome parent, two different pairs.

The gametes, egg cells and pollen cells (sperm), are derived by cell divisions of a special kind, called meiosis (see Chapter 3).One of the essential features of the process of meiosis is that the members of each pair are attracted to each other and come to lie side by side in the nucleus. The meiotic cell divisions are intricate, but the pertinent outcome is the separation of like, or homologous, chromosomes, such that each gamete comes to possess only one member of each pair of chromosomes. Each gamete is said to contain a *haploid* complement of chromosomes, or half the number found in a somatic cell. The latter, in turn, is described as being *diploid* in chromosome number.

The sterility of the first-generation hybrid is now comprehensible. There are simply no homologous chromosomes in the F₁ hybrid. Each chromosome lacks a homologue to act as its pairing partner at meiosis. The process of meiosis in the hybrid is chaotic; the chromosomes move at random into the gametes. The eggs and pollen cells typically contain an odd assortment of chromosomes and are nonfunctional.

Figure 11.2 **Sequence of events** leading to a new, fertile species from two old species by hybridization and polyploidy. The F₁ hybrid plant dervied from a cross of the two paternal species is sterile. The F₁ hybrid may occasionally produce viable gametes when all chromosomes fortuitously enter a gamete during the process of meiosis. The fusion of such gametes leads to a new form of plant, which contains two complete sets of chromosomes (one full set of each of the original parents).

Occasionally, by sheer chance, a few gametes might be produced by the F_1 hybrid that contain all the chromosomes (fig. 11.2). These gametes would be functional, and the fusion of such sex cells would give rise to a plant that contains twice the number of chromosomes that the first-generation hybrid possessed. The plant actually would contain two complete sets of chromosomes; that is, the full diploid complement of chromosomes of each original parent. Such a double diploid is termed a polyploid—more specifically, a *tetraploid.*

The tetraploid hybrid would resemble the first-generation hybrid, but the plant as a whole would be larger and somewhat more robust as a consequence of the increased number of chromosomes. More importantly, the tetraploid hybrid would be fully fertile. The meiotic divisions would be normal, since each chromosome now has a regular pairing partner during meiosis (fig. 11.2). The tetraploid hybrid is a true breeding type; it can perpetuate itself indefinitely. It is, however, reproductively incompatible with its original parental species. If the tetraploid hybrid were to cross with its original parental species, the offspring would be sterile. Hence, the tetraploid hybrid is truly a new distinct species. In but a few generations, we have witnessed essentially the fusion of two old species to form a single derived species.

EXPERIMENTAL VERIFICATION

To return to our wheat story, bread wheat (fig. 11.1) contains the chromosome sets of three diploid species—the *AA* of einkorn wheat, the *BB* of *Aegilops speltoides,* and the *CC* of *Aegilops squarrosa.* (The third set is typically referred to by botanists as the *D* genome. It is an accident of nomenclature that the third genome received the letter *D,* rather than *C.*) Technically, then, the bread wheat contains six sets of chromosomes; it is a *hexaploid.* The relationships of the three major groups of wheat can be shown as follows:

Einkorn = 14 = *AA* = diploid
Emmer = 28 = *AABB* = tetraploid
Bread = 42 = *AABBCC* = hexaploid

Experimental proof was lacking at the time John Percival proposed that the bread wheats originated from hybridization between the emmer wheat and goat grass, followed by chromosome doubling in the hybrid. Verification awaited an effective method of artificially inducing diploid cells to become polyploid. The search for an efficient chemical inducing agent culminated in the discovery in the late 1930s of colchicine, a substance obtained from the roots of the autumn crocus plant. Treatments of diploid plant cells with colchicine result in a high percentage of polyploid nuclei in the treated plant cells. Colchicine acts on the spindle apparatus of a dividing cell and prevents a cell from dividing into two daughter halves. The treated undivided cell contains two sets of daughter chromosomes, which ordinarily would have separated from each other had cell division not been impeded. The cell thus comes to possess twice the usual number of chromosomes.

The experimental production of polyploid cells through the application of colchicine paved the way for studies on wheat by E. S. McFadden and E. Sears of the United States Department of Agriculture. These investigators successfully hybridized a tetraploid species of emmer wheat with the diploid wild grass, *Aegilops squarrosa.* The chromosome number in the hybrids was doubled by treatment with colchicine. The synthetic hexaploid hybrids were similar in characteristics to natural hexaploid species of bread wheat, and they produced functional gametes. At almost the same time, Hitoshi Kihara of Japan obtained a comparable hexaploid wheat species, which spontaneously and naturally had become converted from a sterile hybrid to a fertile hybrid. Kihara's work reinforced the notion that doubling of chromosomes can occur accidentally.

To complete the proof, McFadden and Sears crossed their artificially synthesized hexaploid wheat species with one of the naturally occurring bread wheat *Triticum spelta.* Fully fertile hybrids resulted, removing any doubt that wild grass, a noxious weed, is indeed a parental ancestor of the bread wheats.

COTTON

The story of the cataclysmic evolution of cotton *(Gossypium)* has been partially unraveled by botanical investigators. The evolution of cotton poses some interesting, but unresolved, problems.

Cotton is widely distributed throughout the world and occurs in both the wild and the cultivated state. The cultivated types in the Americas are represented by *Gossypium barbadense,* the prominent cotton of South America, and *Gossypium hirsutum,* grown mainly in Central America and the United States. These two cultivated species of American cotton are of particular interest in that each possess 52 chromosomes (or 26 pairs) in their cells. Thirteen of these pairs of chromosomes are small and resemble those in wild diploid cotton species still found growing in the Americas; the other 13 pairs are large and like those of diploid cotton species native to the Old World. The Old World cotton ancestor contains the *A* chromosome set (or genome); the wild American ancestor possesses the *D* genome. Unquestionably, the present-day American cultivated cotton resulted from a cross between the Old World cotton and the wild American cotton *(AA × DD),* with subsequent natural duplication of the chromosome complement in the hybrid *(AD* to *AADD).* The cultivated American cottons are thus tetraploid progenies of two diploid species. The question as to when and where the two diploid progenitors met and hybridized is a thorny one.

Available evidence indicates that the American tetraploid species arose by hybridization in the coastal valleys of Peru in western South America. Many botanists contend that the Old World diploid parent came from southern Asia, having reached South America by dispersal across the Pacific Ocean. The seeds of cotton, however, are not adapted to transportation over great distances by either water or wind. One authority on cotton, S. C. Harland, has suggested that the Old World parent crossed the Pacific by a land bridge in late Cretaceous times, about 100 million years ago. Most modern geologists dismiss as unreasonable the once popular notion of an ancient Pacific land bridge.

Other botanists, particularly J. B. Hutchinson, R. A. Silow, and S. G. Stephens, have argued that the Asiatic diploid cotton was introduced within historic times by civilized man. Early nomadic peoples from Asia carried the seeds of crop plants across the Pacific to South America. From hybridizations of the transplanted Old World cotton with wild cotton of the valleys of Peru emerged the superior tetraploid plant. This tetraploid cotton was subsequently introduced into Central America and then spread to the United States. The South American tetraploids are no longer reproductively compatible with their northern counterparts; hence, as remarked earlier, the presence today of two distinct species of cultivated cotton in the Americas, *Gossypium barbadense* and *Gossypium hirsutum.*

The reconstruction of the past history of the cultivated American cotton is far from complete. There exists a primitive diploid species of cotton in the southern region of Africa, regarded as the forebear of the Old World diploid cottons. The Asiatic cottons were probably derived from the African type. Recently, there has been speculation that the African species of cotton, rather than the Asiatic species, was the immediate parent of the American cotton. This would be in accord with Wegener's theory of continental drift. In 1912, the Austrian meteorologist and Arctic explorer Alfred Wegener proposed that the earth's continents had once been a huge land mass (a supercontinent) and have reached their present geographical positions by splitting up and drifting across the ocean floors (fig. 11.3). Initially, Wegener's theory was ridiculed. At a meeting of the American Philosophical Society in Philadelphia in the 1920s, the key speaker pronounced Wegener's thesis "utter, damned rot." Today, the idea of continental drift has been revived and has gained a high measure of scientific respectability. Indeed, we now recognize that Wegener was essentially correct.

COTTON AND CONTINENTAL DRIFT

The original single land mass, called *Pangaea,* split and began to move apart about 180 to 200 million

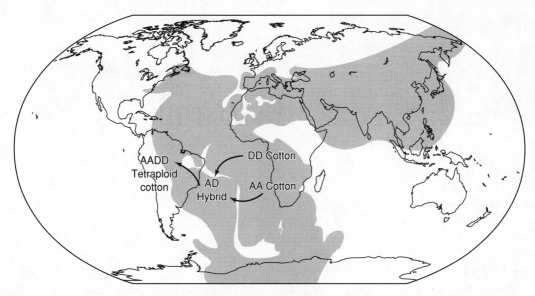

Figure 11.3 **Wegener's theory of continental drift.** The present continents were joined in a single land mass (*Pangaea*) before the start of the Mesozoic era (about 200 million years ago). The hybridization of two African diploid species of cotton to give rise to the American tetraploid species of cotton may have occurred before the continents drifted apart.

years ago. The supercontinent first split into two blocks: *Laurasia* in the Northern Hemisphere, consisting of North America, Europe, and Asia; and *Gondwanaland* in the Southern Hemisphere, made up of Africa, South America, Antarctica, Australia, and India (which then lay far south of Asia). South America began sliding westward about 135 million years ago when a rift appeared between South America and Africa. Antarctica parted from Africa, and India broke free to move northerly toward Asia. North America separated from Europe only about 80 million years ago. It is of interest that India collided with Asia less than 40 million years ago. The collision uplifted the high Tibetan Plateau and raised the Himalayas.

The concept of continental drift serves to explain the present similarity in certain animals and plants in regions of Africa and South America. With respect to cotton, consideration may be given to the possibility that an ancestral diploid cotton species of south Africa (genome *A*) at one time overlapped the distribution of another ancestral

species (genome *D*) in that part of western Africa adjacent to the American continent before continental drift (fig. 11.3). In the area of contact, hybridization between the two diploid species may have occurred naturally, giving rise to the American tetraploid species that became isolated in northern South America at the time of continental separation. This hypothesis, although provocative, limps a little. Africa and South America began to separate 135 million years ago, and perhaps earlier. The fossil history of flowering plants does *not* extend into the geologic past farther back than the Cretaceous period, about 135 million years ago (see chapter 16). It seems unlikely that any of the cotton species could have been in existence prior to the separation of the continents when there is essentially no fossil record of flowering plants before the Cretaceous. The hypothesis could be salvaged by assuming that there was a long pre-Cretaceous evolution of flowering plants, but that pre-Cretaceous conditions were not favorable for the preservation of their fossils.

CHAPTER

12

ADAPTIVE RADIATION

The capacity of a population of organisms to increase its numbers is largely governed by the availability of resources—food, shelter, and space. The available supply of resources in a given environment is limited, whereas the organism's innate ability to multiply is unlimited. A particular environment will soon prove to be inadequate for the number of individuals present. It might thus be expected that some individuals would explore new environments where competition for resources is low. The tendency of individuals to exploit new opportunities is a factor of major significance in the emergence of several new species from an ancestral stock. The successful colonization of previously unoccupied habitats can lead to a rich multitude of diverse species, each better fitted to survive and reproduce under the new conditions than in the ancestral habitat. The spreading of populations into different environments accompanied by divergent adaptive changes of the emigrant populations is called *adaptive radiation*.

GALÁPAGOS ISLANDS

One of the biologically strangest, yet most fascinating, areas of the world is an isolated cluster of islands of volcanic origin in the eastern Pacific, the Galápagos Islands. These islands, which Darwin visited for five weeks in 1835, lie on the equator 600 miles west of Ecuador (fig. 12.1). The islands are composed wholly of volcanic rock; they were never connected with the mainland of South America. The rugged shoreline cliffs are of gray lava and the coastal lowlands are parched, covered with cacti and thorn brushes. In the humid uplands, tall trees flourish in rich black soil.

Giant land-dwelling tortoises still inhabit these islands (see fig. 2.4). After many years of being needlessly slain by pirates and whalers, these remarkable animals now live protected in a sanctuary created in 1959. Still prevalent on the islands are the world's only marine iguana and its inland variety, the land iguana (fig. 12.2). These two species of prehistoric-looking lizards are ancient arrivals from the mainland. The marine forms occur in colonies on the lava shores and swim offshore to feed on seaweed. The land iguana lives on leaves and cactus plants. Cactus fills most of the water needs of the land iguana.

126

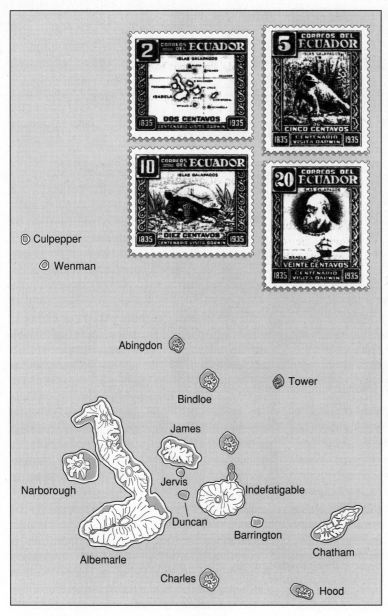

At least 85 different kinds of birds have been recorded on the islands. These include rare cormorants that cannot fly, found only on Narborough Island, and flamingos, which breed on James Island. Of particular interest are the small dark finches. These dusky birds exhibit remarkable variations in the structure of the beak and in feeding habits. The finches afford an outstanding example of adaptive radiation. It was the marked diversity within this small group of birds that gave impetus to Darwin's evolutionary views. Darwin had correctly surmised that the diverse finches were modified descendants of the early, rather homogeneous, colonists of the Galápagos. Our present knowledge of these birds, now appropriately called "Darwin's finches," derives largely from the accomplished work of David Lack at Oxford, who visited the Enchanted Isles in 1938.

Figure 12.1 Galápagos Islands ("Enchanted Isles") in the Pacific Ocean, 600 miles west of Ecuador. Darwin had explored this cluster of isolated islands and found a strange animal life, a "little world within itself." The four stamps shown were issued by Ecuador to commemorate the centenary of Darwin's visit in 1935.

DARWIN'S FINCHES

The present-day assemblage of Darwin's finches descended from small sparrow-like birds that once inhabited the mainland of South America. The ancestors of Darwin's finches were early migrants to the Galápagos Islands and probably the first land birds to reach the islands. These early

Figure 12.2 Land iguana on one of the Galápagos Islands. Despite their horrendous appearance, these bizarre inland lizards are mild, torpid, and vegetarian. They feed on leaves and cactus plants. (Courtesy of American Museum of Natural History.)

beaks. Their beaks are thicker than those of typical flower-eating birds.

All the other species are tree finches, the majority of which feed on insects in the moist forests. One of the most remarkable of these tree dwellers is the woodpecker finch. It possesses a stout, straight beak, but lacks the long tongue characteristic of the true woodpecker. Like a woodpecker, it bores into wood in search of insect larvae, but then it uses a cactus spine or twig to probe out its insect prey from the excavated crevice (fig. 12.4). Equally extraordinary is the warbler finch, which resembles in form and habit the true warbler. Its slender, warbler-like beak is adapted for picking small insects off bushes. Occasionally, like a warbler, it can capture an insect in flight.

colonists have given rise to 14 distinct species, each well adapted to a specific niche (fig. 12.3). Thirteen of these species occur in the Galápagos; one is found in the small isolated Cocos Island, northeast of the Galápagos. Not all 13 species are found on each island.

The most striking differences among the species are in the sizes and shapes of the beak, which are correlated with marked differences in feeding habits. Six of the species are ground finches, with heavy beaks specialized for crushing seeds. Some of the ground finches live mainly on a diet of seeds found on the ground; others feed primarily on the flowers of prickly pear cacti. The cactus eaters possess decurved, flower-probing

FACTORS IN DIVERSIFICATION

No such great diversity of finches can be found on the South American mainland. In the absence of vacant habitats on the continent, the occasion was lacking for the mainland birds to exploit new situations. However, given the unoccupied habitats on the Galápagos Islands, the opportunity presented itself for the invading birds to evolve in new directions. In the absence of competition, the colonists occupied several ecological habitats, the dry lowlands as well as the humid uplands. The finches

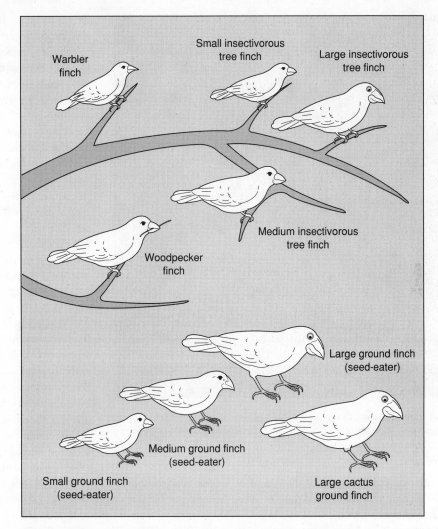

Figure 12.3   Representatives of Darwin's finches. There are 14 species of Darwin's finches, all confined to the Galápagos with the exception of one species that inhabits Cocos Island. Closest to the ancestral stock are the six species of ground finches, primarily seed-eaters. The others evolved into eight species of tree finches, the majority of which feed on insects.

adopted modes of life that ordinarily would not have been opened to them. If true warblers and true woodpeckers had already occupied the islands, it is doubtful that the finches could have evolved into warbler-like and woodpecker-like forms. Thus, *a prime factor promoting adaptive radiation is the absence of competition.*

The emigration of the ancestral finches from the South American mainland was assuredly not conscious or self-directed. The dispersal of birds from the original home was a chance event. Birds (along with other animals and plants) are thought to have been carried westward by the Humboldt Current aboard seaweed mats or driftwood rafts

Figure 12.4 **Tool-using woodpecker finch** *(Camarhynchus pallidus)* prodding for grub with a carefully chosen cactus spine of proper size and shape. (cW/Miguel Castro/Photo Researchers, Inc.)

from the continent. The survivors of the ordeal of raft voyages initiated an evolution of their own on the Galápagos Islands.

The original flock of birds that fortuitously arrived at the islands was but a small sample of the parental population, containing at best a limited portion of the parental gene pool. It is likely that only a small amount of genetic variation was initially available for selection to work on. What evolutionary changes occurred at the outset were due mainly to random survival (genetic drift). However, the phenomenon of drift would become less important as the population increased in size. Selection unquestionably became the main evolutionary agent, molding the individual populations into new shapes by the preservation of new favor-

able mutant or recombination types. More than one island was colonized, and the complete separation of the islands from each other promoted the genetic differentiation of each new local population.

The Galápagos Islands are the tips of enormous volcanoes, most of which rise from 7,000 to 10,000 feet above the floor of the sea. It is of especial interest that five distinct varieties of tortoises occupy regions corresponding to the five principal volcanic tops of the large Island of Albemarle (fig. 12.5). The five different forms of tortoises evolved independently of each other at a time when Albemarle Island was so deeply submerged that only its five major craters were above water. The subsequent lowering of the sea level in the geologic past had the effect of uniting the five volcanoes into a single land mass. Even to this day the five volcanoes are separated from each other by deep valleys. There is no intermingling of the five varieties of tortoises. This intriguing geologic circumstance serves to document the importance of long-term isolation for the genetic differentiation of the separated populations.

Students of evolution had debated whether such unique life forms as the marine and land iguanas could have evolved in the relatively short period of geologic time represented by the present-day Galápagos Islands, less than or equal to three million years old. The contentious question seems to have been resolved by the impressive finding that the chain of Galápagos Islands once

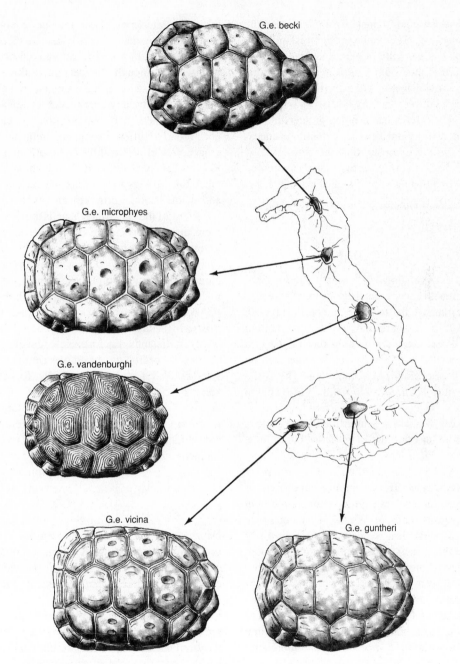

Figure 12.5 Distribution of the five subspecies of the giant land tortoise *(Geochelone elephantopous)* on Albemarle Island. Each of the five subspecies inhabits one of the five major volcanoes on Albemarle. Each subspecies evolved independently on the separate volcanic tops of the island.

included older islands that are now below the ocean's surface. The long-vanished islands are geologically nine million years old and are 370 miles closer to the South American mainland than the present-day islands. Accordingly, some of the inimitable animals, such as the two varieties of iguanas, would have had a much greater interval of time to evolve from a common mainland ancestor than was previously thought. Presumably, many of the life forms on the older, submerged islands found their way to the newer, present-day islands in the chain.

COMPETITIVE EXCLUSION

When two populations of different species are obliged (under experimental conditions) to use a common nutrient, the two will compete with each other for the common resource. Under competition because of identical needs, only one of the two species populations will survive; the other will be eliminated or excluded by competition. This is the principle of *competitive exclusion.* It is also known as *Gause's principle,* after the Russian biologist G. F. Gause, who first demonstrated experimentally that under controlled laboratory conditions, one of two competing species will perish.

In 1934, Gause studied the interactions under carefully controlled culture conditions of two protozoan species, *Paramecium caudatum* and *Paramecium aurelia.* When each species was grown separately in a standard medium in a test tube containing a fixed amount of bacterial food, each species flourished independently. When the two species were placed together in the same culture vessel, however, the growth of *P. caudatum* gradually diminished until the population became eliminated. In enforced competition for the same limited food supply, *P. aurelia* was the more successful species.

Gause enunciated his principle primarily on the basis of observations of "bottle populations" in the laboratory. The competition experiments reveal that two species populations cannot exist together if they are competing for precisely the same limited

resource. Alternatively, if two species in nature were to occupy the same habitat, the expectation is that each would have different ecological requirements, even though the degree of difference is slight. Ecologists have demonstrated the validity of this view. In virtually every natural situation carefully examined, two co-inhabiting species have been found to differ in some requirement. The heterogeneous resources of the environment in a given locality are typically partitioned among the co-inhabiting species to minimize direct competition and enable the two (or more) species to coexist.

Most of the Galápagos Islands are occupied by more than one species of finch. On islands where several species of finch exist together, we find that each species is adapted to a different ecological niche. The three common species of ground finch—small *(Geospiza fuliginosa),* medium *(Geospiza fortis),* and large *(Geospiza magnitrostris)*—occur together in the coastal lowlands of several islands. Each species, however, is specialized in feeding on a seed of a certain size. The small-beaked *Geospiza fulginosa,* for example, feeds on small grass seeds, whereas the large-beaked *Geospiza magnitrostis* eats large, hard fruits. Different species, with different food requirements, can thus exist together in an environment with varied food resources.

COEXISTENCE

Gause's principle may be stated in the following form: No two species with identical requirements can continue to exist together. However, it is exceedingly unlikely that two species in nature would have *exactly* the same requirements for food and habitat. The sum total of environmental requirements for a species to thrive and reproduce has been termed "the niche" of that species population. The term *niche,* as ecologists use it, is more than simply the physical space that the species population occupies. It is essentially the way of life peculiar to a given species: its structural adaptations, physiological responses, and behavior within its habitat. Experience has shown that the likelihood of two species having identical niches is almost nil.

Direct evidence of the process of competition between species in nature is difficult to obtain. Observable competitive interaction is a relatively fleeting stage in the relation of two species populations. What the ecologist observes is the end result of competitive contact, when the actual or potential competitors have become differentially specialized to exploit different components of a local environment and accordingly live side by side. The outcome, then, of incipient competitive interaction is the *avoidance* of competition through differential specialization, or in the terminology of ecologists, through *niche diversification.*

Indirect arguments have been used to support the view that two closely related species populations come to exploit different ecological niches in the same locality after a beginning of competitive interaction. We may envision a situation in which two closely related species, with almost similar ecological requirements, expand their geographic ranges and meet in a common habitat. It may be presumed that each of the two species populations has appreciable genetic variability and that the resources in the common habitat are varied. The two species populations would initially compete for suitable ecological niches in the new common habitat. However, we can expect that the members of the two species will ultimately become so different in structure and behavior that each species will become specialized to use different components of the environment. In other words, if genetic differences in morphological and behavioral characteristics tend at first to reduce competition between the two species, then

subsequently natural selection will act to augment the differences between the two competing species. It is especially noteworthy that the differences between the two species become more pronounced as a consequence of selection *reducing* competition rather than *intensifying* competition.

A striking case of niche diversification has been described by David Lack for certain species of *Geospiza* in the Galápagos. Where two species occur together on an island, there is a conspicuous difference in the size of the beaks and food habits. Where either species exists alone on other islands, the beak is adapted to exploit more than one food resource. Thus, on Tower Island (fig. 12.6), the cactus-feeding *Geospiza conirostris* and the large seed-eating ground finch *Geospiza magnirostris* live side by side. The former species occurs also on Hood Island, but the latter species is absent, presumably having failed to invade or reach the island. In the absence of competition, *Geospiza conirostris*

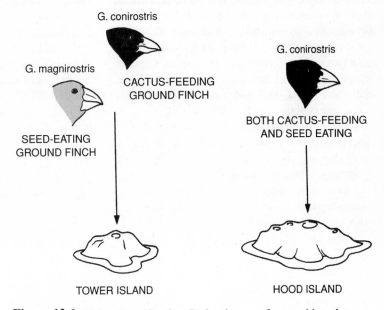

Figure 12.6 Niche diversification. In the absence of competition, the ground finch *Geospiza conirostris* on Hood Island has evolved a large beak that enables it to exploit a variety of food items (seed and cacti). On Tower Island, the same ground finch *Geospiza conirostris* has evolved a specialized beak adapted solely for cactus feeding, thus reducing or avoiding competition for food resources with the seed-eating ground finch *Geospiza magnirostris*. The latter species does not exist on Hood Island.

on Hood Island has evolved a larger beak that is adapted to feeding on both cactus and seeds. The sharp separation in beak size of the two species on Tower Island is understandable if we assume that competition initially fostered the differentiation of the beaks to permit each of the two species to adapt to a limited or restricted range of the available food resources. Each species is now genetically specialized in food habits, and competition between the two is now avoided.

There are many examples to illustrate how two or more species avoid competition. Among warblers that inhabit the spruce forests of Maine, each species confines its feeding to a particular region of the spruce tree. The myrtle warbler, for example, preys on insects at the base of the spruce tree, whereas the Blackburnian warbler prefers those insects on the exterior leaves of the top of the tree. They can exist together because they use different resources of the same tree.

We may conclude by stating, paradoxically, that competition between two species populations achieves the avoidance or reduction of further competition, and not an intensification. In natural populations, coexistence of two species, rather than competitive exclusion, is the general rule. The end result is that each species is part of a highly organized community in which each plays a constructive, or stabilizing, role. Nevertheless, there is one species—*Homo sapiens*—that seems unable to live in harmony with other species. Our destruction of natural habitats has led to the extinction of several species of wildlife (for example, the passenger pigeon) and the reduction in rank of many others

(for example, the bison). The survival of numerous species continues to be threatened, among them the grizzly bear, the ivory-billed woodpecker, and the whooping crane. In the United States alone, approximately 100 species of wildlife were listed recently as being in danger of becoming extinct. Even more dramatic, of the United States' 20,000 native plants, about 4,200 are threatened with extinction.

CLASSIFICATION

The original ancestral stock of finches on the Galápagos diverged along several different paths. The pattern of divergence is reflected in the biologists' scheme of classification of organisms. All the finches are related to one another, but the various species of ground finches obviously are more related by descent to one another than to the members of the tree-finch assemblage. As a measure of evolutionary affinities, the ground finches are grouped together in one genus *(Geospiza)* and the tree finches are clustered in another genus *(Camarhynchus)*. The different lineages of finches are portrayed in figure 12.7. It should be clear that our classification scheme is nothing more than an expression of evolutionary relationships between groups of organisms.

In the next chapter we shall see how adaptive radiation on a much larger scale than that which occurred in the finches led to the origin of radically new assemblages of organisms, distinguishable by the taxonomist as orders and classes.

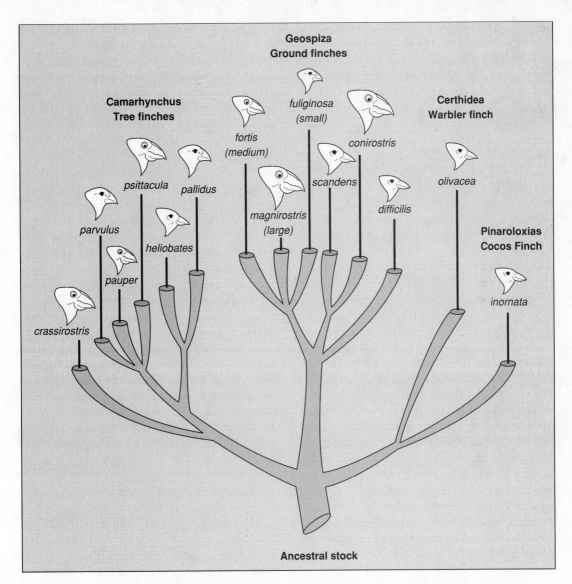

Figure 12.7 Evolutionary tree of Darwin's finches, graphically expressing what is known or surmised as to the degree of relationship or kinship between the different species of finches. Darwin's finches evolved from a common stock and are represented today by 14 species. The species are assembled in four genera, of which two (Certhidea and Pinaroloxias) contain only a single species each. The species associated together in a genus have significant common attributes, judged to denote evolutionary affinities. (Based on the findings of David Lack.)

13

MAJOR ADAPTIVE RADIATIONS

The diversity of Darwin's finches had its beginning when migrants from the mainland successfully invaded the variety of vacant habitats on the Galápagos Islands. Adaptive radiation, such as manifested by Darwin's finches, has been imitated repeatedly by different forms of life. Organisms throughout the ages have seized new opportunities open to them by the absence of competitors and have diverged in the new environments. The habitats available to Darwin's finches were certainly few in comparison with the enormous range of ecological habitats in the world. The larger the region and the more diverse the environmental conditions, the greater the variety of life.

Approximately 400 million years ago, during a period of history that geologists call the *Devonian,* the vast areas of land were monotonously barren of animal life. Save for rare creatures like scorpions and millipedes, animal life of those distant years was confined to the water. The seas were crowded with invertebrate animals of varied kinds. The fresh and salt waters contained a highly diversified and abundant assemblage of cartilaginous and bony fishes. The vacant terrestri-

al regions were not to remain unoccupied for long. From one of the many groups of fishes inhabiting the pools and swamps in the Devonian period emerged the first land vertebrate. The initial modest step onto land started the vertebrates on their conquest of all available terrestrial habitats. The origin and diversification of the backboned land dwellers is one of the most striking examples of adaptive radiation on a large scale.

INVASION OF LAND

Prominent among the numerous Devonian aquatic forms were the lobe-finned fishes, the Crossopterygii, who possessed the ability to gulp air when they rose to the surface. These ancient air-breathing fishes represent the stock from which the first land vertebrates, the amphibians, were derived (fig. 13.1). The factors that led these ancestral lobe-finned fishes to venture onto land are unknown. The impelling force might have been population pressure or simply the inherent tendency of individuals, particularly the young, to disperse.

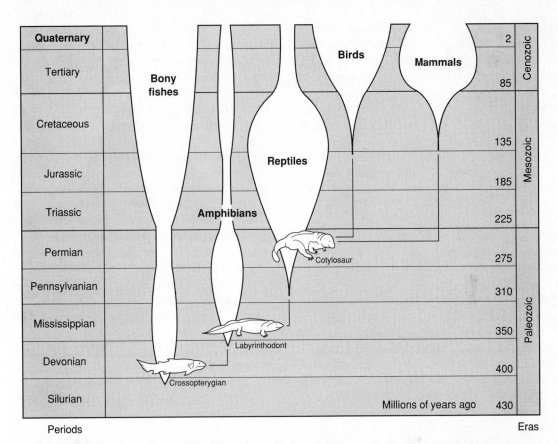

Figure 13.1 Evolution of land vertebrates in the geologic past. From air breathing, lobe-finned fishes (crossopterygians) emerged the first four-footed land inhabitants, the amphibians. Primitive amphibians (labyrinthodonts) gave rise to the reptiles, the first vertebrates to become firmly established on land. The birds and mammals owe their origin to an early reptilian stock (cotylosaurs). An important biological principle reveals itself: each new vertebrate group did not arise from highly developed or advanced members of the ancestral group but rather from early primitive forms near the base of the ancestral stock. The thickness of the various branches provides a rough measure of the comparative abundance of the five vertebrate groups during geologic history. The Devonian period is often called the *Age of Fishes;* the Mississippian and Pennsylvanian periods (frequently lumped together as the Carboniferous period) are referred to as the *Age of Amphibians;* the Mesozoic era is the grand *Age of Reptiles;* and the Cenozoic era is the *Age of Mammals.*

The prominent paleontologist A. S. Romer advanced the conjecture that the crossopterygians were forced to crawl on dry land on those occasions when the pools they inhabited became foul, stagnant, or completely dry. There is convincing geological evidence that the Devonian years were marked by excessive seasonal droughts. It is not unimaginable that the water in some pools periodically evaporated. The suggestion, then, is that the crossopterygians wriggled out of stagnant and shrinking water holes onto land to seek pools in which water still remained. Thus, paradoxically, the first actual movements on land might not have been associated at all with an attempt to abandon aquatic existence, but rather to retain it.

The foregoing hypothesis is admittedly speculative. However, the fact remains that these crossopterygians who emerged on land, though crudely adapted for terrestrial existence, did not encounter any competitors who could immediately spell doom to their awkward initial trial on land. We should note that the lobe-finned fishes did possess certain capacities that would prove to be important under the new conditions of life. Evolutionists speak of such potential adaptive characters as *preadaptations*.

The preadaptations of the lobe-finned fishes included primitive membranous lungs and internal nostrils, both of which are important for atmospheric breathing (fig. 13.2). It should be understood that such preadapted characters were not favorably selected with a view to their possible use in some future mode of life. There is no foresight or design in the selection process. A structure does not evolve in advance of impending events. Nor do mutational changes occur in anticipation of some new environmental condition. A trait is selected only when it imparts an advantage to the organism in its immediate environment. Accordingly, lungs in the crossopterygians did not evolve with conscious reference toward a possible future land life, but only because such a structure was important, if not essential, to the survival of these air-breathing fishes in their immediate surroundings. In essence, the selection process does not foresee environments yet to come. The expression "preadaptation" simply signifies that a structure evolves under one set of environmental circumstances and becomes modified later for a totally different function.

The crossopterygians did not, of course, possess typical amphibian limbs. However, their lateral fins contained fleshy lobes, within which were bony elements that were basically comparable to those of a limb of a terrestrial vertebrate. Figure 13.3 shows a restoration of a widespread Devonian form, *Eusthenopteron,* in which the lateral fins had developed into stout, muscular paddles. It bears emphasizing that these primitive amphibians arose from a population of air-breathing fishes that fortuitously had the minimal preadaptive features to survive in the vacant (terrestrial) ecological niche.

Before the close of the Devonian period, the transition from fish to amphibian had been completed. The early land-living amphibians were

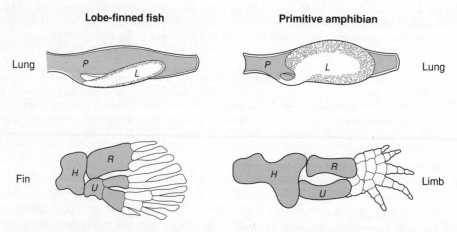

Figure 13.2 The evolution of lungs and limbs. The main elements of the limb—humerus *(H)*, radius *(R),* and ulna *(U)*—are clearly comparable in a lobe-finned fish and an early amphibian. The irregular outer branches of the fish limb are transformed into a pattern of digits in a land animal. The lungs represent outpocketings from the underside of the pharynx *(P)* of the digestive tract. The foldings of the internal walls of the lungs in land animals improve the efficiency of the lungs.

slim-bodied with fishlike tails but possessed limbs capable of locomotion on land. The four-footed amphibians flourished in the humid swamps of Mississippian and Pennsylvanian times but never did become completely adapted for existence on land. All the ancient amphibians, such as *Diplovertebron* (fig. 13.3), spent much of their lives in water, and their modern descendants—the salamanders, newts, frogs, and toads—must return to water to deposit their eggs. Thus, the amphibians were the first vertebrates to colonize land but were, and still are, only partially adapted for terrestrial life.

CONQUEST OF LAND

From the amphibians emerged the reptiles, true terrestrial forms. The appearance of a shell-covered egg, which can be laid on land, freed the reptile from dependence on water. The elimination of a water-dwelling stage was a significant evolutionary advance. The first primitive reptile most likely arose during Carboniferous times, but the fossil beds of this period have yet to reveal the appropriate reptilian ancestor. The ancestral reptile probably possessed the body proportions and well-developed limbs of the more advanced terrestrial forms of amphibians, such as *Seymouria* (fig. 13.4). *Seymouria* is not a reptile, but its skeletal features suggest terrestrial habits. This advanced amphibian may have been a descendant of an earlier amphibian group in the lower Carboniferous period, a group that was ancestral to the reptiles.

The terrestrial egg-laying habit evolved very early in reptilian evolution. A key feature of reptiles (and higher vertebrates) is the *amniotic egg* (fig. 13.5). The egg of the reptile (or bird) contains a large amount of nourishing yolk. Moreover, development of the embryo takes place entirely within a thick shell. These circumstances call for a special provision whereby food derived from the yolk and oxygen obtained from an external source can be made accessible to all parts of the developing embryo. The embryo itself constructs a complex system of membranes, known

Figure 13.3 Stages in the transition of the lobe-finned fishes into amphibians, as reconstructed by W. K. Gregory and painted by Francis Lee Jaques. *Bottom,* the primitive Devonian air-breathing crossopterygian, *Eusthenopteron,* floundering on the stream bank with its muscular, paddlelike fins. *Top,* the Pennsylvanian tailed amphibian, *Diplovertebron,* with limbs capable of true locomotion on land. Much of the life of this early tetrapod was spent in the water. (Courtesy of American Museum of Natural History.)

Figure 13.4 **Seymouria,** an advanced amphibian with reptilian body proportions (short trunk and well-developed limbs). *Seymouria* is known only from Permian rocks (275 million years ago), long after the reptiles had appeared. It may represent a relict of an earlier amphibian group that was ancestral to the reptiles. (Photo L. Botin & R. Logan Courtesy Dept. of Library Services American Museum of Natural History.)

as *extraembryonic membranes,* which serve for protection, nutrition, and respiration. These are the *amnion, chorion, yolk sac,* and *allantois.*

When fully developed, the amnion is a thin membrane loosely enclosing the embryo. It does not fit the embryo snugly. The space between the amnion and the embryo, the amniotic cavity, is filled with a watery fluid, the amniotic fluid. The fluid is a cushion that protects the embryo from mechanical impacts and, at the same time, allows it freedom of movement. The chemical composition of the amniotic fluid resembles the chemical make-up of blood. Thus, during its development, the embryo is bathed by a fluid that is compatible in its chemical nature with the embryo's blood. An interesting fact is that for the embryos of birds and reptiles, as well as for the human fetus, the amniotic fluid resembles seawater in chemical composition, in terms of elements (ions), such as sodium, potas-

sium, calcium, and magnesium. After approximately 400 million years of evolution on land, vertebrates still carry essentially the chemical composition of seawater in their fluids—amniotic fluid, blood, and other body fluids. This has been interpreted as indicating that life originated in the sea and that the balance of salts in various body fluids did not change very much in subsequent evolution. The amnion has been picturesquely characterized as a sort of private aquarium in which the embryos of land-living vertebrates recapitulate the water-living mode of existence of their remote ancestors.

As figure 13.5 shows, the enormous mass of yolk has been enclosed by an internal circular membrane, the yolk sac. This membrane, which has formed by growing over the yolk, is attached to the embryonic body by a narrow stalk. The yolk sac is highly vascular, its many blood vessels communicating with the blood channels of the embryo proper. The blood circulating through the vessels of the yolk sac carries dissolved yolk materials to all parts of the embryo, thus making the yolk available for chemical activities and growth in all regions.

From the hind region of the embryonic body, a sac bulges out on the underside and pushes its way between the yolk sac and chorion. This sac is the allantois (fig. 13.5). An appreciable part of the blood vessels lies close to the inner surface of the shell. The porous shell permits the ready exchange of respiratory gases between the external air and the internal blood. The allantoic sac serves also as a receptacle for urinary wastes. Waste fluids excreted by the embryonic kidneys pass into the cavity of the allantois. The allantois thus has both respiratory and excretory functions. Finally, there is an outermost investing membrane, the chorion, which is abundantly supplied with blood vessels. The embryo depends, in part, on this vascularized enveloping membrane for carrying on gaseous exchange with the outer air through the porous shell.

The amnion, chorion, allantois, and yolk sac in the human embryo are similar in essence to these extraembryonic membranes in the reptiles (and birds). Curiously, the yolk sac does develop in the human embryo in spite of the absence of any appreciable amount of yolk. The human yolk sac,

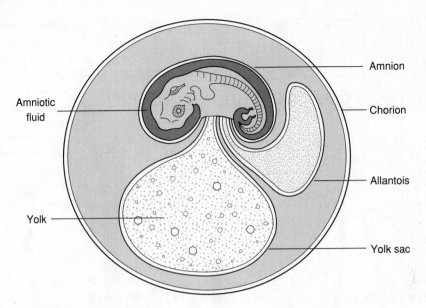

Figure 13.5 The amniotic egg. The developing embryo is enclosed by the membranous *amnion* and cushioned by amniotic fluid. The large reserve supply of food (yolk) is contained within the *yolk sac*. The sacklike *allantois* serves as a receptacle for the embryo's waste products. The outermost, vascularized enveloping membrane is the *chorion*.

however, remains small and largely functionless. There is also no elaborate development of the allantois in the human embryo; the allantoic sac never becomes more than a rudimentary tube of minute size. Nevertheless, the very appearance of the yolk sac and allantois in the human embryo is one of the strongest pieces of evidence documenting the evolutionary relationships among the widely different kinds of vertebrates. To the student of evolution this means that the mammals, including human beings, are descended from animals that reproduced by means of externally laid eggs rich in yolk.

ADAPTIVE RADIATION OF REPTILES

With the perfection of the amniotic egg, the reptiles exploited the wide expanses of land areas. The ancestral reptilian stock initiated one of the most spectacular adaptive radiations in life's history. The reptiles endured as dominant land animals of the earth for well over 100 million years. The Mesozoic era, during which the reptiles thrived, is often referred to as the "Age of Reptiles."

Figure 13.6 reveals the variety of reptiles that blossomed from the basal stock, the cotylosaurs. The dinosaurs were by far the most awe inspiring and famous. They reigned over the land until the close of the Mesozoic era when they became extinct. The dinosaurs were remarkably diverse; they varied in size, bodily form, and habits. Some of the dinosaurs were carnivorous, such as the huge *Tyrannosaurus,* whereas others were vegetarians, such as the feeble-toothed but ponderous *Brontosaurus.* The prodigious body of *Brontosaurus* weighed 30 tons and measured nearly 70 feet in length. Not all dinosaurs were immense; some were no bigger than chickens. Some dinosaurs strode on two feet; others had reverted to four. The exceedingly long necks of certain dinosaurs were adaptations for feeding on the foliage of tall coniferous trees.

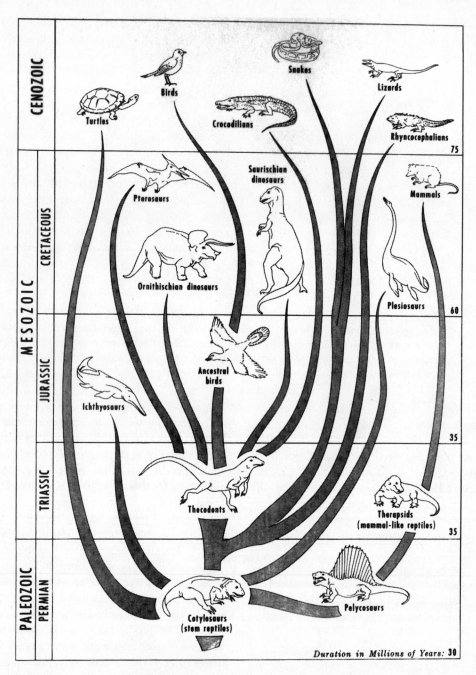

Figure 13.6 Adaptive radiation of reptiles. A vast horde of reptiles came into existence from the basal stock, the cotylosaurs, at the beginning of the Mesozoic era, roughly 225 million years ago. This matchless assemblage of reptiles was triumphant for a duration well over 100 million years. Then, before the close of the Mesozoic era, the great majority of reptiles passed into oblivion.

The dinosaurs were descended from the thecodonts—slender, fast-running lizardlike creatures. In fact, there were two great groups of dinosaurs, the Saurischia and the Ornithischia (fig. 13.6), which evolved independently from two different lines of the thecodonts. The thecodonts gave rise also to bizarre reptiles that took to the air, the pterosaurs. These "dragons of the air" possessed highly expansive wings and disproportionately short bodies. The winged pterosaurs succumbed before the end of the Mesozoic era. Another independent branch of the thecodonts led to eminently more successful flyers, the birds. The origin of birds from reptiles is revealed by the celebrated *Archaeopteryx,* a Jurassic form (fig. 13.6) that was essentially an air-borne lizard. This feathered creature possessed a slender, lizardlike tail and a scaly head equipped with reptilian teeth. Some authors contend that birds descended from reptiles that were already warm-blooded (endothermic) rather than, as often assumed, cold-blooded (ectothermic). Under this view, the endothermic birds are merely aerial extensions of a terrestrial endothermic stock.

Certain reptiles returned to water. The streamlined, dolphinlike ichthyosaurs and the long-necked, short-bodied plesiosaurs were marine, fish-eating reptiles. The ichthyosaurs were proficient swimmers; their limbs were fin-like and their tails were forked. The plesiosaurs were efficient predators, capable of swinging their heads 40 feet from side to side and seizing fish in their long, sharp teeth. These aquatic reptiles breathed by means of lungs; they did not redevelop the gills of their very distant fish ancestors. Indeed, it is axiomatic that a structure once lost in the long course of evolution cannot be regained. This is the doctrine of irreversibility of evolution, or *Dollo's Law,* after Louis Dollo, the eminent Belgian paleontologist to whom the principle is ascribed. The reversal of the long evolutionary path from lungs to gills would demand the implausible precise retrieval of an untold number of steps.

Among the early reptiles present before Mesozoic days were the pelycosaurs, notable for their peculiar sail-like extensions of the back (fig. 13.7). The function of the gaudy sail is unknown, but it should not be thought that this structural feature was merely ornamental or useless. As we have emphasized, traits of organisms were selected for their adaptive utility. It may be

Figure 13.7 Dimetrodon, one of the pelycosaurs that flourished during Permian times. The gaudy "sail" may have served as a heat-regulating device. (From a restoration by Charles R. Knight, courtesy of American Museum of Natural History.)

that the pelycosauran sail was a functional device to achieve some degree of heat regulation. Be that as it may, the pelycosaurs gave rise to an important group of reptiles, the therapsids. These mammal-like forms bridged the structural gap between the reptiles and the mammals.

EXTINCTION AND REPLACEMENT

The history of the reptiles attests to the ultimate fate of many groups of organism—*extinction.* The reptilian dynasty collapsed before the close of the Mesozoic era. Of the vast host of Mesozoic reptiles, relatively few have survived to modern times; the ones that have include the lizards, snakes, crocodiles, and turtles. The famed land dinosaurs, the great marine plesiosaurs and ichthyosaurs, and the flying pterosaurs all became extinct. The cause of the decline and death of the tremendous array of reptiles has been linked in time to a large extraterrestrial impact (comet or meteor) in the vicinity of the Yucatan peninsula. Whatever radical environmental changes this impact created, dinosaurs and many other reptiles were unable to adapt to the altered environmental conditions.

Whatever may be the cause of mass extinctions, the fact remains that as one group of organisms recedes and dies out completely, another group spreads and evolves. The decline of the reptiles provided evolutionary opportunities for the birds and the mammals. The vacancies in the habitats could then be occupied by these warm-blooded vertebrates. Small and inconspicuous during the Mesozoic era, the mammals arose to unquestionable dominance during the Cenozoic era, which began approximately 65 million years ago. The extent to which the mammals contributed to the decline of the dinosaur assemblage is unanswerable. Our knowledge of evolutionary history suggests that the ascendancy of the mammals was the result rather than the cause of the fall of the great reptiles.

In mammals, a major evolutionary advance was the development of the *placenta,* an intimate apposition of maternal and fetal tissues. The placenta provides optimum conditions for the growth of the fetus—protection, a continuous supply of oxygen and food, an efficient system for removing waste products, and the transmission of antibodies. Of paramount significance, mammals have added a higher order of parental care of the young. The newborn is relatively helpless and obtains nourishment in the form of milk through special mammary glands. The most primitive mammals are the monotremes (duckbill platypus and spiny anteater), which lay reptile-like yolky eggs. Yet, the monotremes nourish their young with milk secreted by modified sweat glands.

The mammals diversified into marine forms (for example, the whale, dolphin, seal, and walrus), fossorial forms living underground (for example, the mole), flying and gliding animals (for example, the bat and the flying squirrel), and cursorial types well-adapted for running (for example, the horse). The mammalian appendages are highly differentiated for the different modes of life. An important lesson may be drawn from the variety of specialized appendages. Superficially, there is scant resemblance between the human arm, the flipper of a whale, and the wing of a bat. And yet, a close comparison of the skeletal elements (fig. 13.8) shows that the structural design, bone for bone, is basically the same. The differences are mainly in the relative lengths of the component bones. In the forelimb of the bat, for instance, the metacarpals and phalanges (except those of the thumb) are greatly elongated. Although highly modified, the bones of the bat's wing are not fundamentally different from those of other mammals. The conclusion is inescapable that the limb bones of the human, the bat, and the whale are modifications of a common ancestral pattern. The facts admit no other logical interpretation. Indeed, as seen in figure 13.8, the forelimbs of all tetrapod vertebrates exhibit a unity of anatomical pattern intelligible only on the basis of common inheritance. The corresponding limb bones of tetrapod vertebrates are said to be *homologous,* since they are structurally identical with those in the common ancestor. Homology implies evolutionary *divergence* from a common ancestor. In contrast, the wing of a bird and the wing of a butterfly are *analogous;* both are used for flight, but they are built on an entirely different structural plan and are of independent origin *(convergence).*

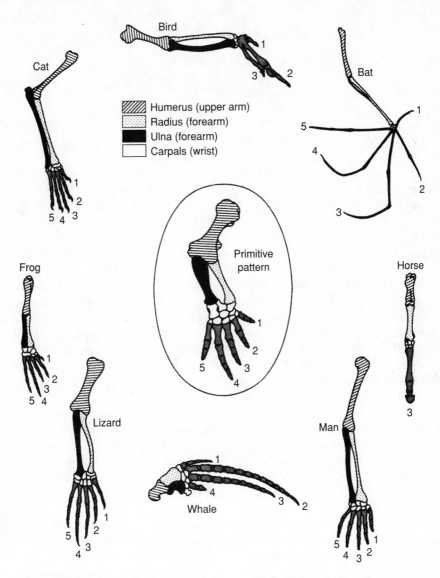

Figure 13.8 Varied forelimbs of vertebrates, all of which are built on the same structural plan. The best explanation for the fundamentally similar framework of bones is that humans and all other vertebrates share a common ancestry. Homologous bones are shaded appropriately. The number of phalanges in each digit is indicated by a numeral, beginning with the first digit (thumb).

14

ORIGIN OF LIFE

The polemics surrounding Darwin's *The Origin of Species* was matched by another monumental controversy of that day, a controversy regarding the unbelievable proposition that germs (bacteria) cause disease. The father of the germ theory was Louis Pasteur, the chemist who was led into the study of germs through problems associated with the fermentation of wine. In the early 19th century, the notion of spontaneous generation was still flourishing—that is, most people believed that living things could originate from lifeless matter. In 1862, Pasteur provided proof that new life can come only from preexisting life. Following years of bitter wrangling with skeptics, Pasteur finally laid to rest the age-old belief that life could appear out of nowhere.

Since Pasteur's time, countless generations of students have been taught not to believe in spontaneous generation. However, Pasteur's experiments revealed only that life cannot arise spontaneously *under conditions that exist on earth today*. Conditions on the primeval earth billions of years ago were assuredly different from present conditions, and the first form of life, or

self-duplicating particle, did arise spontaneously from chemical, inanimate substances.

We are unable to witness the unique molecular events that led to the first form of life ages ago. The scientist's reconstruction of life's origin is largely circumstantial, but persuasive.

PRIMITIVE EARTH

The view that life emerged through a long and gradual process of chemical evolution was first convincingly set forth by the Russian biochemist Alexander I. Oparin, in 1924, in an enthralling booklet entitled *The Origin of Life*. The transformation of lifeless chemicals into living matter extended over some one billion years. Such a transformation, as Oparin points out, is no longer possible today. If by pure chance a living particle approaching that of the first form of life should now appear, it would be rapidly decomposed by oxygen of the air or quickly destroyed by the countless microorganisms presently populating the earth.

Our earth is reliably estimated to be 4.6 billion years old. It was formerly thought that the earth originated as a fiery mass that was torn away from the sun. Astronomers now generally acknowledge that the earth (like other planets in the solar system) condensed out of a swirling cloud of gas surrounding the primitive sun. The atmosphere of the pristine earth was quite unlike our present atmosphere. Oxygen in the free gaseous state was virtually absent; it was bound in water and in metallic oxides on surface rocks and particles. Accordingly, any complex organic compound that arose during this early time would not be subject to degradation by free oxygen. Moreover, there was no layer of ozone to absorb the stark ultraviolet rays from the sun, which would be lethal to modern animal life.

The early gas cloud was especially rich in hydrogen. The hydrogen (H_2) of the primordial earth chemically united with carbon to form methane (CH_4), with nitrogen to form ammonia (NH_3), and with oxygen to form water vapor (H_2O). Thus, the early atmosphere had a strongly reducing (nonoxygenic) character, containing primarily hydrogen, methane, ammonia, and water. The atmospheric water vapor condensed into drops and fell as rain; the rains eroded the rocks and washed minerals (such as chlorides and phosphates) into the seas. The stage was set for the combination of the varied chemical elements. Chemicals from the atmosphere mixed and reacted with those in the waters to form a wealth of hydrocarbons (that is, compounds of hydrogen and carbon). Water, hydrocarbons, and ammonia are the raw materials of amino acids, which, in turn, are the building blocks for the larger protein molecules. Thus, in the primitive seas, amino acids accumulated in considerable quantities and became linked together to form proteins.

Complex carbon compounds, such as proteins, are termed organic because they are made by living organisms. Our present-day green plants use the energy of sunlight to synthesize organic compounds from simple molecules. What, then, was the energy source in the primitive earth, and how was synthesis of organic compounds effected in the absence of living things? It is generally held

that ultraviolet rays from the sun, electrical discharges, such as lightning, and intense dry heat from volcanic activity furnished the energy to join the simple carbon compounds and nitrogenous substances into amino acids. Is there a valid basis for such a widely accepted view?

EXPERIMENTAL SYNTHESIS OF ORGANIC COMPOUNDS

In the early days of chemistry, it was believed that organic compounds could be produced only by living organisms. But in 1828, Friedrich Wöhler succeeded in manufacturing the organic compound *urea* under artificial conditions in the laboratory. Since Wöhler's discovery, a large variety of organic chemicals (amino acids, monosaccharides, purines, and vitamins) formerly produced only in organisms have been artificially synthesized.

In 1953, Stanley Miller, then at the University of Chicago and a student of Nobel Laureate Harold Urey, synthesized organic compounds under conditions resembling the primitive atmosphere of the earth. He passed electrical sparks through a mixture of hydrogen, water, ammonia, and methane (fig. 14.1). The electrical discharges duplicated the effects of violent electrical storms in the primitive universe. In the laboratory, the four simple inorganic molecules interacted, after a mere week, to form several kinds of amino acids, among them alanine, glycine, aspartic acid, and glutamic acid. Hydrogen cyanide and cyanoacetylene were also identified among the products, and those compounds can serve as intermediates in the formation of nitrogenous bases (purines and pyrimidines) of nucleic acids. Miller's instructive experiment has been successfully repeated by a number of investigators; amino acids and nucleotides can also be generated by irradiating a similar mixture of gases with ultraviolet light.

It is very likely that proteins were randomly produced abiotically on the primitive earth. However, life itself apparently did not begin with

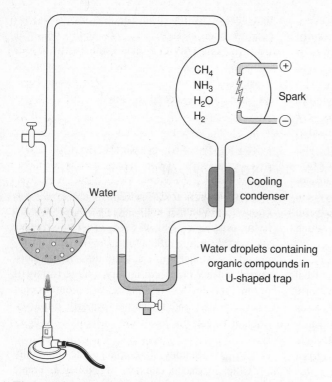

Figure 14.1 Stanley Miller's apparatus simulating conditions of the primordial earth. The upper chamber contains a mixture of gases (methane, ammonia, and hydrogen gas plus water vapor). An electrical discharge (simulating lightning) is passed through the vaporized mixture. The mixture is cooled by a water condenser, and any complex molecules formed in the atmosphere chamber are dissolved in the water droplets. The organic molecules accumulate in the U-shaped trap, where samples are withdrawn for analysis.

University and Thomas Cech of the University of Colorado, who independently discovered that RNA can act as an enzyme *(ribozyme)*. Thus, RNA can catalyze its own replication as well as pass on hereditary information.

LIFE'S BEGINNINGS

It is clear that organic compounds can be formed without the intervention of living organisms. Thus, it appears likely that the sea of the primitive earth spontaneously accumulated a rich mixture of organic molecules. In the absence of living organisms, the organic compounds would have been stable and would persist for countless years. The sea became a sort of dilute organic soup (an aquatic Garden of Eden), in which the molecules collided and associated to form larger molecules (polymers) of increasing levels of complexity. Proteins capable of catalysis, or enzymatic activity, had to evolve, and nucleic acid molecules capable of self-replication must also have developed.

The living cell is an orderly system of chemical reactions (directed by enzymes) that has the ability to reproduce. It seems reasonable that the machinery for self-replication (that is, self-duplicating nucleic acids) evolved before the development of the metabolic (enzymatic) machinery of the cell. Whereas amino acids are the building blocks of proteins, nucleotides are the basic units of nucleic acids. The linear polynucleotides occur in nature in the form of deoxyribonucleic acid (DNA) or ribonucleic acid (RNA). Polynucleotides have the unique property of specifying the sequence of nucleotides in a new molecule by acting as templates for the assembly of the nucleotides. Since preferential binding occurs between pairs of

a self-sustaining network of proteins. Polypeptide chains of amino acids are not self-complementary and, accordingly, cannot serve as templates for their own replication. On the other hand, nucleic acids are capable of self-replication. As discussed below, ribonucleic acid (RNA) probably was present before the emergence of protein synthesis. Moreover, RNA evidently preceded DNA in the evolution of life. The perspective that RNA arose before proteinaceous enzymes (catalysts) gained support from the work of Sydney Altman of Yale nucleotides (cytosine coupled with guanine and adenine coupled with uracil), new daughter polynucleotide molecules can arise in which the nucleotide sequences are complementary to the parent molecules. A second round of copying, in which the complementary strand serves as a template, restores the original sequence. Thus, the original random sequence is multiplied many times.

Self-replicating polynucleotides (nucleic acids) slowly became established in the primordial earth 3.5 billion years ago. In the absence of proteinaceous enzymes in the prebiotic environment, less efficient catalysts in the form of minerals or metal ions would be sufficient to promote the reactivity of the nucleotide precursors. More importantly, however, RNA itself can act as a catalyst. The linear RNA molecule can fold up to form complex surfaces that catalyze specific reactions. Accordingly, RNA molecules have enzymatic properties in addition to their well-known properties as templates. It is highly probable that RNA guided the primordial synthesis of protein (fig. 14.2). The nucleotides of RNA (not DNA) were probably the first carriers of genetic information. RNA established a primitive genetic code for ordering amino acids into proteins. It warrants emphasis that the genetic code—the translation of nucleotide sequences into amino acids sequences—became established at a very early stage of organic evolution. The universality of the genetic code in present-day organisms attests to the early origin of the code.

We may presume that errors continually occurred in the copying process, and that the sequence of nucleotides in the original polynucleotide molecule became altered on numerous occasions. Large numbers of polynucleotide variations were undoubtedly maladapted. Just as organisms today compete for available resources, molecules with different nucleotide sequences competed for the available nucleotide precursors in promoting copies of themselves. Any new mutant sequence with a replication rate higher than the antecedent sequence would be the

"fittest" and would prevail. Natural selection (differential reproduction) thus operates on populations of molecules as it does on populations of organisms.

A selectively advantageous RNA molecule would be one that directs the synthesis of a protein that accelerates the replication of that particular RNA. However, a protein specified by a particular variant of RNA could not foster the reproduction of that kind of RNA unless it were restrained in the immediate vicinity of the RNA. The free diffusion of proteins could be forestalled if some form of a compartment evolved to enclose or circumscribe the specific protein made by a particular RNA (fig.14.2). All present-day cells have a limiting (plasma) membrane composed principally of phospholipids. It is not implausible that the first cell was formed when polarized films of phospholipids formed soaplike bubbles enclosing aggregations of other complex macromolecules. Specifically, a membranous structure (plasma membrane) arose that encircled a self-replicating aggregation of RNA and protein molecules. Once bounded by a limiting membrane, a given RNA molecule could be assured of propagating its own special protein molecule.

At some later stage in the evolutionary process, DNA took the place of RNA as the repository of genetic information. In other words, the genetic code was stored in a more stable DNA double helix. Today, RNA molecules serve essentially their primal function: directing protein synthesis. Thus, in modern cells, genetic information is stored in DNA, transcribed into RNA, and translated into protein. An endless permutation of nucleic acids and proteins has fostered an enormous richness of life.

We may assume that the first living systems drew upon the wealth of organic materials in the sea broth. Organisms that are nutritionally dependent on their environment for ready-made organic substances are called *heterotrophs* (Greek, *hetero*, "other" and *trophos,* "one that feeds"). The primitive one-celled heterotroph probably had little more than a few genes, a few proteins, and a

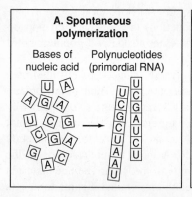

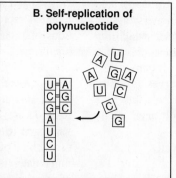

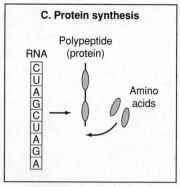

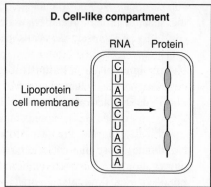

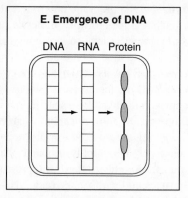

Figure 14.2 Postulated stages in the origin of living cells. *(A)* Polynucleotides (primordial RNA) emerge from the spontaneous association of nitrogenous bases (adenine, cytosine, guanine, and uracil). *(B)* RNA has enzymatic properties and catalyzes its own replication. *(C)* Information encoded in the base sequences of RNA specifies the amino acid sequence in a polypeptide chain (protein). *(D)* A spontaneously associated lipoprotein serves as a cell membrane enclosing RNA and its protein product. *(E)* DNA replaces RNA as the repository of hereditary information.

limiting cell membrane. The heterotrophs multiplied rapidly in an environment with a copious supply of dissolved organic substances. However, the ancient heterotrophs could survive only as long as the existing store of organic molecules lasted. Eventually, living systems evolved the ability to synthesize their own organic requirements from simple inorganic substances. In the course of time, *autotrophs* (*auto,* meaning "self") arose, which were able to manufacture organic nutrients from simpler molecules.

The primitive autotroph would require a whole array of enzymes to direct a multistep chain of reactions involved in the synthesis of a protein.

However, it would be too much to expect that all the necessary enzymes evolved at the same time. Norman H. Horowitz of the California Institute of Technology has ingeniously proposed that the chain of steps in a complicated chemical pathway evolved *backward.*

RETROGRADE EVOLUTION

Enzymes required in a biosynthetic pathway normally appear in a sequence. Horowitz proposed that an organism might, by successive mutations, acquire these enzymes in reverse order. In other words, evolution began with the end product of the pathway and worked backward, one step at a time, toward the beginning of the reaction chain. The pathway became constructed in its entirety when the organism had tapped from its surroundings all of the intermediate precursors of the reaction.

Let us assume that an organic compound, O, is synthesized through the following steps, where A through D represent precursors:

$$A \rightarrow B \rightarrow C \rightarrow D \rightarrow O$$

The first primitive heterotroph, lacking synthetic ability, would require the presence of O in its environment. This essential organic chemical, continually being used by the heterotroph, would eventually become rare, if not exhausted. If a mutation occurred that endowed the heterotroph with the ability to catalyze the reaction from D to O, then the organism would no longer be dependent on the availability of O in the environment. Indeed, in an environment where O had become depleted, the new mutant would have survival (selective) advantage over the ancestral type and would replace it. As the new mutant reproduced, D in turn would become scarce. Another mutation might occur in the mutant, converting it to a still newer form capable of catalyzing the reaction C to D. At this point, both the first mutant and the newly arisen mutant could live together in a close mutual (symbiotic) relationship, establishing the first two-gene system.

The intimate two-gene combination would be able to survive and reproduce in the presence of C and D. As other compounds in the synthetic pathway become progressively rarer, additional mutations for their synthesis would be favorably selected. Ultimately, a multigenic system would evolve capable of directing the synthesis of O from inorganic substance A by way of intermediate products B, C, D. At first glance, it might appear that an unreasonable number of mutational events has been invoked. But it should be recalled that mutations continually occur in living organisms. In fact, the capacity for mutation might be regarded as a cardinal property of life.

AUTOTROPHIC EXISTENCE

The first simple autotroph arose in an anaerobic world, one in which little, if any, free oxygen was available. The primitive autotrophs obtained their energy from the relatively inefficient process of fermentation (the breakdown of organic compounds in the absence of oxygen). Thus, the early fermentative autotrophs were much like our present-day anaerobic bacteria and yeast. The metabolic processes of the anaerobic autotrophs resulted in the liberation of large amounts of carbon dioxide into the atmosphere. Once this occurred, the way was paved for the evolution of organisms that could use carbon dioxide as the sole source of carbon in synthesizing organic compounds and could use sunlight as the sole source of energy. Such organisms would be the photosynthetic cells.

Early photosynthetic cells probably split hydrogen-containing compounds, such as hydrogen sulfide. In other words, hydrogen sulfide was cleaved into hydrogen and sulfur. (This is still done today by sulfur bacteria.) The hydrogen was used by the cell to synthesize organic compounds, and the sulfur was released as a waste product (as evidenced by the earth's great sulfur deposits). In time, the process of photosynthesis was refined so that water served as the source of hydrogen. The result was the release of oxygen as a waste product. At this stage, free oxygen became established for the first time in the atmosphere.

The first organisms to use water as the hydrogen source in photosynthesis were the blue-green algae. Since blue-green algae were active photosynthesizers, atmospheric oxygen accumulated in increasing amounts. Oxygen would be toxic to anaerobic bacteria. Many primitive anaerobic bacteria, incapable of adapting to free oxygen, remained in portions of the environment that were anaerobic, such as sulfur springs and oxygen-free muds. However, new kinds of bacteria arose that were capable of utilizing the free oxygen. Today, there are bacterial types that are anaerobic as well as aerobic.

The earth's atmosphere gradually changed from a reducing, or hydrogen-rich, atmosphere to an oxidizing, or oxygen-rich, atmosphere. Geologic evidence indicates that free oxygen began to accumulate in the atmosphere about 2 billion years ago. The rising levels of atmospheric oxygen set the stage for the appearance of one-celled eukaryotic organisms, which arose at least 1 billion years ago (fig. 14.3). Then, within the comparatively short span of the last 600 million years, the one-celled eukaryote evolved in various directions to give rise to a wealth of multicellular life forms inhabiting the earth.

PROKARYOTIC AND EUKARYOTIC CELLS

Some of the simplest cells in nature include bacteria and cyanobacteria ("blue-green algae"), which have been classified as *prokaryotes*. The Greek expression for nucleus is *karyon;* hence, prokaryote means "before a nucleus." Prokaryotic cells have no nuclear membrane by which the hereditary materials (DNA) are set apart from the cytoplasm and they lack specialized cytoplasmic bodies (organelles), such as mitochondria and choloroplasts. In a bacterial cell, the DNA forms a simple loop that is attached to the inside of the cell's membrane. In the absence of mitochondria, the enzymatic machinery in bacteria for converting energy into usable form is found in the plasma membrane. The replication of a bacterial cell occurs rapidly by the simple division of the cell into two. One bacterial cell can divide every 20 minutes and can leave 4 billion descendants within 11 hours.

In contrast to the prokaryotes, the cells of the morphologically more complex plants and animals, or *eukaryotes,* have a distinct nuclear membrane that encloses discrete DNA-containing chromosomes. The eukaryotes include not only multicellular animals and plants but also unicellular organisms, such as protozoa, fungi, and many algae. Eukaryotic cells also have an elaborate system of membrane-bound cytoplasmic organelles. The enzymes that are responsible for releasing large amounts of energy are packed inside the mitochondria. In the eukaryotic plant cell, a prominent organelle is the chloroplast, the photosynthesizing agent that converts light energy into chemical energy. One of the most intriguing speculations is that mitochondria arose not by a gradual evolutionary process but abruptly in an unusually striking manner.

ORGANELLES AND EVOLUTION

Mitochondria have a number of interesting properties that suggest they were once free-living, or independent, bacteria-like organisms. Mitochondria have small amounts of their own DNA distinct from nuclear DNA. Mitochondrial DNA exists as a loop-shaped molecule like the DNA of bacteria. In the early 1970s, Lynn Margulis, then at Boston University, enunciated the provocative hypothesis that mitochondria may have been derived from primitive aerobic bacteria that were engulfed by predatory organisms, probably fermentative bacteria, destined to become eukaryotic (fig.14.4). The predatory hosts became dependent on their enslaved mitochondria, and the latter, in turn, became dependent on their hosts. Thus, the association was of mutual advantage to the predatory cell and the engulfed prey. Such a close association, or partnership, is called *symbiosis.* In essence, the mitochondria (formerly oxygen-respiring bacteria) established permanent residence within the hosts. The predatory hosts became the first eukaryotic cells.

Eukaryotic plant cells contain both mitochondria and chloroplasts. Chloroplasts, like mitochondria, have their own unique DNA and the associated protein-synthesizing machinery. We can now speculate that those predatory cells that engulfed but did not digest the aerobic bacterial cells became the eukaryotic animal cells. Those predators that captured both the photosynthetic blue-green algae cells and the aerobic bacteria evolved into eukaryotic plant cells (fig. 14.4). Accordingly, the engulfed prey became permanent symbiotic residents—either as mitochondria or chloroplasts—within the predatory cell. The startling idea that present-day mitochondria and chloroplasts were descendants of ancestral aerobic bacteria and blue-green algae, respectively, has

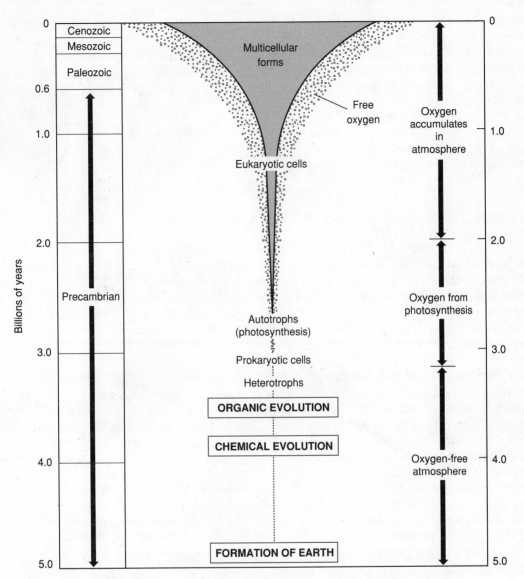

Figure 14.3 Major events in the history of life. Chemical evolution preceded biological evolution. The first self-duplicating life forms were heterotrophs. Oxygen began to accumulate in the atmosphere with the appearance of photosynthetic autotrophs. Rising levels of atmospheric oxygen were associated with the emergence of eukaryotes. The last 600 million years have witnessed a great diversity of life in an oxygen-rich atmosphere.

gained widespread acceptance in the scientific community through the years.

Although the relatively abrupt origin of mitochondria and choloroplasts in eukaryotes is now generally accepted, it is likely that the evolution of the nucleus and endoplasmic reticulum proceeded through a series of gradual changes. Continual modification of the surface membrane of a prokaryotic cell may have been the means by which the nucleus and endoplasmic reticulum of a eukaryotic cell

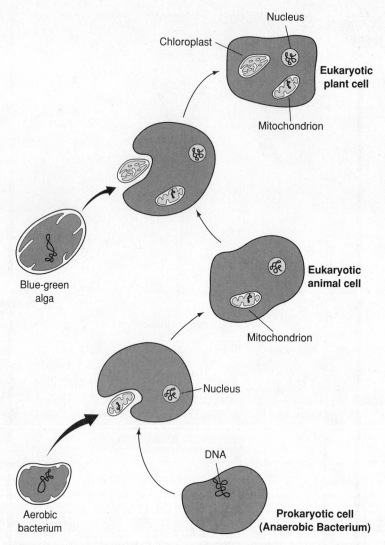

Figure 14.4 Evolution of eukaryotes through symbiosis. The eukaryotic animal cell may have arisen through a symbiotic relationship between a prokaryotic cell and an aerobic bacterium. The eukaryotic plant cell may have originated by a comparable symbiotic relationship between a blue-green alga and a eukaryotic animal cell.

originated. In other words, the nucleus and endoplasmic reticulum may have evolved by the invagination, or drawing inward, of the surface membrane of a primitive cell. Figure 14.5 shows the postulated mechanism for the origin of the nucleus and endoplasmic reticulum from the cell surface membrane.

CLASSIFICATION OF ORGANISMS

Aristotle (384–322 B.C.), the great philosopher and naturalist, aptly declared that Nature is marvelous in each and all her ways. Although Aristotle did not view different kinds of organisms as being related

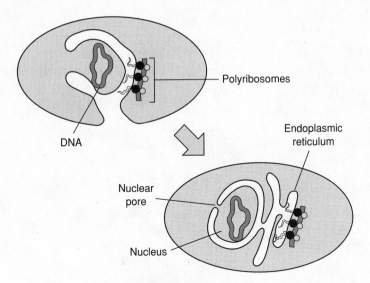

Figure 14.5 Hypothetical scheme of the origin of the nucleus and
endoplasmic reticulum of a eukaryotic cell by the invagination, or
drawing inward, of the surface membrane.

by descent, he arranged all living things in an
ascending ladder with humans at the top. A formal
scheme of classification was developed by the
Swedish naturalist Karl von Linné (1707–1778),
whose name generally appears in latinized form,
Carolus Linnaeus. Life was divided by Linnaeus
into two grand kingdoms, *Animals* and *Vegetables,*
broadly defined as follows:

ANIMALS adorn the exterior parts of the
earth, respire, and generate eggs; are impelled to
action by hunger, congeneric affections, and pain;
and by preying on other animals and vegetables,
restrain within proper proportion the numbers of
both.

They are bodies *organized,* and *have life, sensation,* and *the power of locomotion.*

VEGETABLES clothe the surface with ver-
dure, imbibe nourishment through bibulous roots,
breathe by quivering leaves, celebrate their nup-
tials in a genial metamorphosis, and continue their
kind by the dispersion of seed within prescribed
limits.

They are bodies *organized,* and *have life* and
not sensation.

Linnaeus was convinced that all species of
animals and plants were fixed, unchanging entities,
and that "there are just as many species as there
were created in the beginning." Modern taxonomists have departed completely from Linnaeus'
stand. Present-day taxonomy is based on the
proposition of evolutionary change. The modern
classification scheme is an expression of the evolutionary relationships among groups of organisms. All systems of classification are imperfect to
the extent that our knowledge is imperfect. The
classification scheme is continually being modified as new information on the evolutionary rela-
tions of organisms comes to light.

In 1969, R. H. Whittaker suggested a radical
departure from the traditional two-kingdom system. He grouped organisms into five separate
kingdoms (fig.14.6). Whittaker's system takes into
consideration the fundamental differences
between prokaryotic and eukaryotic levels of
organization and stresses the principal modes of
nutrition—photosynthetic, absorptive, and inges-
tive. The five kingdoms are the Monera (unicellular, prokaryotic organisms), Prostista (unicellular,

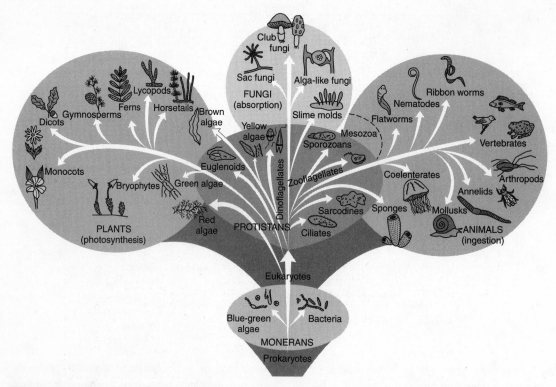

Figure 14.6 The five-kingdom classification. Primitive unicellular prokaryotes (kingdom *Monera*) gave rise to more complex single-celled eukaryotes (kindom *Protista*). The divergence of the Protistans led to the three kingdoms of *Plantae, Fungi,* and *Animalia.*
(Original design contributed by Stuart S. Bamforth, Tulane University.)

eukaryotic organisms), Plantae (multicellular higher algae and green plants), Fungi (multinucleate plantlike organisms lacking photosynthetic pigments), and Animalia (multicellular animals). Whittaker's scheme was favorably received and adopted by most taxonomists for many years. We have become accustomed to dividing all life forms into two major categories based on cell types, the prokaryotes (bacteria and blue-green algae) and the eukaryotes.

In 1997, Carl R. Woese at the University of Illinois brought to light an unusual group of one-celled organisms so different from other forms of life, including bacteria, that he placed them in a separate domain of life. These unusual microbes, which flourish in harsh conditions, were initially

called *archaebacteria* but are now referred to as *archaea* (Woese himself dropped the "bacteria"). Sometimes thought of as "extremophiles," the strange group of archaea consists of bacteria-like organisms that generate methane (known as methanogens), thrive in salt (halophiles), tolerate acids (acidophiles), endure extreme heat (thermophiles and hyperthermophiles), and prosper in strongly alkaline conditions (alkaliphiles).

Woese proposed a novel universal phylogenetic tree (fig. 14.7). Instead of five major kingdoms, there are three broad domains: Bacteria, Archaea, and Eukarya. In Woese's scheme, the Archaea and the Bacteria appear first and diverge from their common ancestor relatively soon in the history of life. Later, the Eukarya branches off

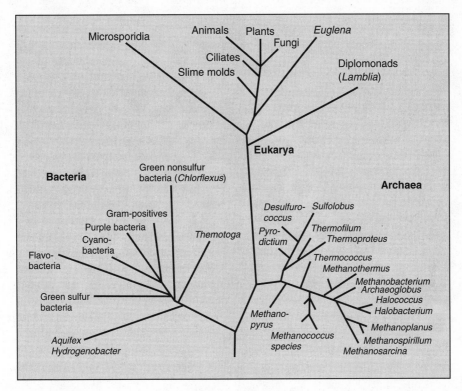

Figure 14.7 The three broad domains of Carl Woese's classification scheme: Bacteria, Archaea, and Eukarya. The predominant forms of life appear to be one-celled.

(From *Science Magazine*, vol 276. No. 5313, 2 May 1997, p. 701.)

from the Archaea. The one-celled organisms seem to dominate the tree of life. The multicellular organisms (Eukarya) appear as a mere twig in a great microbial world!

Woese used the tools of molecular biology in reassigning life into three broad domains. He relied heavily on the nucleotide sequences found in ribosomal RNA, which are highly conserved in all organisms and would aptly serve to chronicle the past evolutionary history. Woese subsequently undertook the sequencing of the complete genome of one of the archaea, *Methanococcus jannaschii*. His work, reported in 1996, showed that *M. jannaschii* has a host of genes unlike those of any

organism known. Specifically, nearly 56 percent of the gene sequences are completely different from any so far described from either bacteria or eukaryotes. Of the 44 percent of the genome that does match other life forms, some are recognizable as bacteria-like while others, particularly those controlling transcription and translation, resemble eukaryotic genes. The Archaea apparently are a rather unique group. Indeed, some investigators have now suggested that life did not begin in an inviting warm sea bath, but rather in notoriously hot environments, perhaps like Yellowstone's sulfurous hot springs or a deep-sea vent.

15

MOLECULAR EVOLUTION

One of the dramatic breakthroughs in the 1950s was an understanding of the interrelations of protein structure and the genetic information carried in deoxyribonucleic acid (DNA). There emerged the fundamental tenet that the gene (a linear array of nitrogenous bases in the DNA) codes for the precise sequence of amino acids that compose a protein. There was also the realization that gene mutation often involves no more than an alteration of a single base in the DNA sequence that specifies the amino acid. Such a single base alteration can result in the substitution in the protein molecule of one amino acid for another. This interferes with the activity of the protein. This is vividly exemplified by the aberrant hemoglobin of sickle-cell anemia, in which a highly localized change in one of the bases of the DNA molecule leads to the substitution of only one amino acid residue in the hemoglobin molecule.

Over evolutionary time, DNA sequences diverge and proteins change in composition. The amino acid sequences of proteins have gradually been modified with time, yielding the arrays that are found today in extant species. Comparisons of the amino acid compositions in present-day organisms enable us to infer the molecular events that occurred in the past. The more distant in the past that an ancestral stock diverged into two present-day species, the more changes will be evident in the amino acid sequences of the proteins of the contemporary species. Viewed another way, the number of amino acid modifications in the lines of descent can be used as a measure of time since the divergence of the two species from a common ancestor. In essence, protein molecules incorporate a record of their evolutionary history that can be just as informative as the fossil record.

NEUTRAL THEORY OF MOLECULAR EVOLUTION

Electrophoretic methods of detecting differences in the amino acid sequences of proteins in organisms have disclosed a wealth of amino acid substitutions in the course of evolution. Ardent advocates of Darwinian selection, popularly called *selectionists,* maintain that a mutated form of a gene responsible for an altered amino acid must pass the stringent test of natural selection. Selectionists assert that a mutant allele is favored only when an amino acid substitution confers an advantage to the organism. In the late 1960s, the Japanese population geneticist Motto Kimura championed the view that the majority of amino acid substitutions are not likely to provide either an evolutionary advantage or disadvantage and are preserved by sheer chance. In essence, Kimura's postulate is that most amino acid replacements are not fostered by natural selection but result from *neutral* mutations that are fixed by *random drift*. The frequency of a new neutral mutant allele fluctuates over time, increasing or decreasing fortuitously. Many of these alleles will be lost purely by chance in a few generations. But an occasional neutral mutant will spread through the population to reach fixation, or a frequency of 100 percent.

Neutralists do not discount a role for natural selection at the molecular level. The role of selection is to protect a molecule from *deleterious* mutations. Indeed, most newly arising mutations are deleterious and are weeded out by natural selection. Natural selection also, in its positive role, permits the incorporation of *beneficial* mutant alleles. Beneficial mutant alleles are, of course, evolutionarily very important, but they are rare. Accordingly, the majority of variant alleles witnessed today at the molecular level are likely to have been selectively neutral mutations that have been fixed by random drift.

In response to critics, Kimura and his colleague, Tomoko Ohta, a gifted population geneticist in her own right, continually sharpened the neutral thesis of molecular evolution. Kimura and Ohta theorized that new neutral genes insinuate themselves at a steady, or constant, rate in a given lineage. Over periods of millions of years, new forms of genes randomly establish themselves at a steady pace in the population and may themselves be gradually replaced. This turnover of alleles, governed purely by chance, is responsible for the coexistence of the multitude of amino acid alternatives in a protein.

With this theoretical background, we shall turn to the experimental studies on varied proteins and examine the data that provide support for the rather unorthodox doctrine that molecular evolution is not primarily the result of the accumulation of advantageous mutations, but rather the steady accumulation of selectively neutral mutations that do not impair protein function.

CYTOCHROME *c*

Cytochrome *c* is an ancient, evolutionarily conservative molecule that serves as an essential enzyme in respiration. It is a relatively small protein of a chain length slightly over 100 amino acid residues, which apparently exists in all eukaryotic organisms. No other protein has been so fully analyzed for so many different organisms. This ubiquitous protein has been extracted, purified, and analyzed in more than 40 eukaryotic species. It has 104 amino acids in vertebrates and a few more in certain species lower in the phylogenetic tree. Cytochrome *c* performs the same vital function in all organisms.

Approximately a third of the amino acid residues of cytochrome *c* have not varied at all throughout time. In particular, the same amino acids have been found in 35 positions in all organisms tested, from molds to humans. Evidently, certain amino acids at specific positions are irreplaceable or *invariant*. Any substitutions at these invariant sites are likely to interfere with the chemical integrity or folding properties of the protein molecule. Mutational changes at these sites are lethal and are rejected by natural selection. No such evolutionary constraints have been placed on the acceptance of chance alterations at other sites. Amino acid substitutions are tolerated elsewhere since they preserve the essential chemical properties of the molecule. It is these tolerable amino acid substitutions that are promulgated by Kimura's selectively neutral mutations.

The number of amino acid replacements in cytochrome *c* of several species are compared in table 15.1. First, it is evident that the cytochromes

TABLE 15.1	Evolution of Cytochrome *c**

A. Comparison between Different Organisms

Organisms	Number of Variant Amino Acid Residues
Cow and sheep	0
Cow and whale	2
Horse and cow	3
Rabbit and pig	4
Horse and rabbit	5
Whale and kangaroo	6
Rabbit and pigeon	7
Shark and tuna	19
Tuna and fruit fly	21
Tuna and moth	28
Yeast and mold	38
Wheat germ and yeast	40
Moth and yeast	44

B. Comparison with Human Cytochrome *c*

Organisms	Number of Variant Amino Acid Residues
Chimpanzee	0
Rhesus monkey	1
Rabbit	9
Cow	10
Kangaroo	10
Duck	11
Pigeon	12
Rattlesnake	14
Bullfrog	18
Tuna	20
Fruit fly	24
Moth	26
Wheat germ	37
Mold (*Neurospora*)	40
Baker's yeast	42

*Compiled from studies by R. E. Dickerson.

of closely related vertebrates differ in only a few residues, or not at all. There is no difference at all in the composition of the amino acid residues between humans and the chimpanzee. Cytochrome *c* of the rhesus monkey differs from the human only at position 66 of the molecule, where threonine is present instead of isoleucine. Secondly, the greater the phylogenetic differences, the greater

the likelihood that the cytochrome *c* compositions differ. This is vividly apparent when one compares a vertebrate with an insect (tuna and moth), or the human with Baker's yeast.

MOLECULAR CLOCK

Amino acid substitutions accumulate at fairly steady rates over long periods of evolutionary time. The uniform rate is understandable because it is governed primarily by the rate of mutation without the restraint of natural selection. The clocklike rate of molecular evolution was first detected by the Nobel laureate Linus Pauling and Emile Zuckerkandl at the California Institute of Technology, in their now classical analysis of the hemoglobin molecule. The differences in amino acid sequences between two living species can be used to estimate the period of time that has elapsed since the two species diverged from a common ancestor. The number of allelic substitutions generated per time serves as a *molecular clock*. Figure 15.1 shows the constant rate at which the cytochrome *c* gene has evolved.

The molecular clock does not tick at the same rate for all protein molecules. Indeed, the rates of amino acid substitutions are appreciably different in several protein molecules that have been extensively studied. Cytochrome *c* has evolved very slowly in comparison to the fibrinopeptide A, in which the rate of amino acid substitution has been exceedingly rapid. In terms of actual numbers, the rates of amino acid substitutions for cytochrome *c* and fibrinopeptide A are 0.3×10^{-9} and 9.0×10^{-9}, respectively. It is likely that evolutionary change at the molecular level is slow where there are strong functional constraints and faster where changes are least likely to disrupt protein function.

Those who adhere to the neutralist position point with satisfaction to the analysis of the fibrinopeptides. These are two 20-residue fragments (A and B) that are cleaved from the protein fibrinogen when it is activated to form fibrin in a blood clot. Among several mammalian species, there is an extraordinarily high level of amino acid

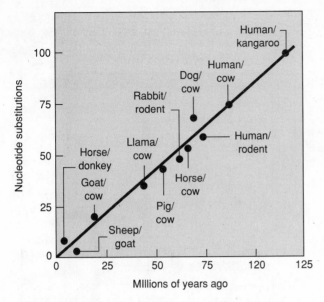

Figure 15.1 The number of differences ("substitutions") in nucleotide sequences in the cytochrome *c* genes of pairs of organisms is plotted against the time evolutionists estimate had elapsed since the pairs of organisms diverged. The resulting straight line signifies a constant rate of nucleotide substitutions, which is interpreted as indicating that cytochrome *c* is evolving at a steady, clocklike pace in all lineages.

(Reprinted, by permission, from H. P. Raven and G. B. Johnson, *Biology*, 4th ed. Dubuque, IA: WCB/McGraw-Hill, 1996, fig. 21.6.)

substitutions in fibrinopeptide A. This peptide apparently functions equally well with numerous different amino acid sequences. Indeed, virtually any amino acid change appears to have been acceptable. It has been argued persuasively that the mutations responsible for the many different amino acids must be selectively neutral. Since the fibrinopeptides have little known function except to permit the activation of fibrinogen by being discarded, they evidently can tolerate any amino acid change.

There are a few proteins in which nearly the entire amino acid sequence is critical for protein function. In such cases, it would be difficult to deny the strong role of natural selection. Certain histones that bind to DNA in the nucleus have been highly conserved in their amino acid sequence. For example, histone IV from such divergent organisms as the pea plant and the calf

differ in only 2 of 102 amino acid residues. The amino acid substitution rate has been calculated at a paltry 0.006×10^{-9}. Nearly all amino acid substitutions would impair the activity of histone IV, and any substitutions, save a few, that may have appeared in the evolutionary past were most likely eliminated by natural selection. Since histone IV plays a vital role in the expression of the hereditary information encoded in DNA, it is hardly surprising that this protein would be closely specified and scrutinized.

If proteins evolve at a constant rate, then the data on amino acid substitutions can be used to construct—or reconstruct—phylogenetic relationships among organisms. As a generality, most molecular phylogenetic trees are in accord with the tempo of evolution based on fossil findings. There are a few notable exceptions. Indeed, in one instance, the molecular data have dramatically modified our understanding of taxonomic affinities and evolutionary time scales. In the mid-1960s, Vincent M. Sarich, working in the laboratory of Allan C. Wilson at the University of California at Berkeley, compared the albumin molecules of humans and African apes. He calculated that the divergence time between humans and African apes was only a scant five million years. At the time of Sarich's pronouncement, the prevailing view among anthropologists was that humans and apes diverged at least 15 million years ago. This view was based on a 15-year-old fossil finding, *Ramapithecus* (see chapter 17), who was purported to be one of the earlier members of the human family. Ironically, the human status of *Ramapithecus* was undermined by new fossil finds. *Ramapithecus* has been dethroned as a forebear of hominids and is presently considered an ancestor of the orangutan. This paved the way for the acceptance of Sarich's thesis that the road to humankind began 5 million years ago, or at least less than 10 million years ago. Equally important, as seen in figure 15.2, we have

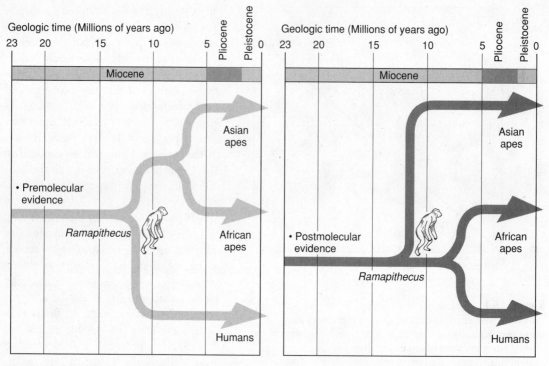

Figure 15.2 On the basis of anatomical features and fossil findings (the *premolecular evidence; left illustration*), the three great apes (chimpanzee, gorilla, and orangutan) were judged to be more closely related to each other than to humans. Genetic analyses *(postmolecular evidence; right illustration)* indicate that the African apes (chimpanzee and gorilla) are more closely related to humans than to the Asian apes (orangutan).

come to the realization that the African apes (chimpanzee and gorilla) are more closely related to humans than they are to Asian apes (orangutan).

HEMOGLOBIN

Human hemoglobin has provided an ideal model for inferring the molecular events that have occurred during evolution. The oxygen-carrying molecule of the red blood cell is made up of four polypeptide chains. The hemoglobin molecule of an adult human is composed of two identical alpha chains and two identical beta chains. Each alpha chain is a linear array of 141 amino acid residues; each beta chain has 146 amino acid residues. In the amino acid sequence of the alpha chain, there is only one difference in the composition of the amino acid residues between humans and the

gorilla. This contrasts sharply with differences of 19 amino acid residues in the alpha chain between humans and the pig, or differences of 26 amino acid units between humans and the rabbit. The striking molecular similarity in the hemoglobin of humans and apes holds up as well for the beta chain.

The syntheses of the alpha and beta chains of normal adult hemoglobin (designated Hb A) are specified by two pairs of genes at different loci. This state of affairs is shown in figure 15.3, where α and β denote the respective genes that control the syntheses of the polypeptide chains. The illustration also reveals that adult hemoglobin is made up of two fractions: hemoglobin A (Hb A), which is the largest component, and hemoglobin A_2 (Hb A_2), which constitutes about 2.5 percent. Hemoglobin A_2 consists of two alpha chains, which are identical to those in hemoglobin A, and two delta chains.

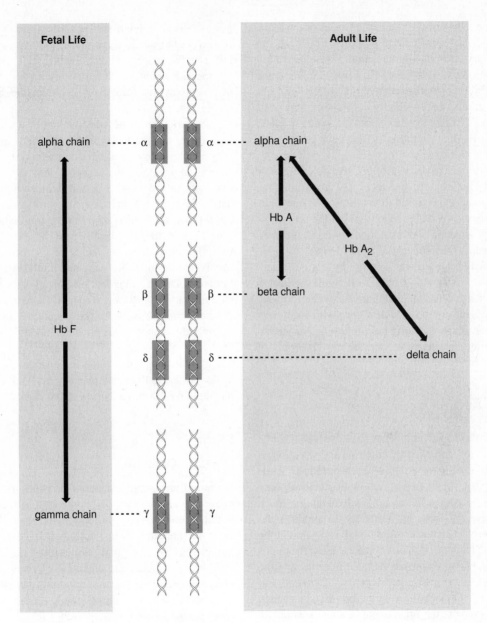

Figure 15.3 Genetic control of hemoglobin by four loci. The beta locus and the delta locus
are closely linked in the same chromosome.

A different hemoglobin is present during fetal
life. The alpha chain of fetal hemoglobin (Hb F) is
chemically identical to the alpha chain of adult
hemoglobin (Hb A) and under control by the same

gene. However, in place of the beta chain, fetal
hemoglobin contains a chemically different
gamma chain that is governed by a separate gene
(γ). Fetal hemoglobin allows increased amounts of

oxygen to be transported to the tissues of the fetus. Fetal hemoglobin carries as much as 30 percent more oxygen at low tensions than does adult hemoglobin. The level of Hb F is high during the last trimester of pregnancy but then falls off gradually until there is virtually none six months postnatally. As the synthesis of the gamma chain decreases, the synthesis of the beta chain takes over.

The locus *(δ)* controlling the delta chain is closely adjacent to the locus *(β)* governing the beta chain. This has led to the thesis that one locus arose as a copy of the other through the process of gene duplication and mutational changes. A redundant copy of a functional gene derived by gene duplication can escape scrutiny by natural selection. While sheltered from selection, the duplicated locus can accumulate mutations that were formerly rejected and emerge as a new gene with a different function than the original gene. As discussed below, duplicated genes are the raw materials of molecular evolution.

GENE DUPLICATION

The process by which genes have evolved through duplication, followed by mutation and selection, is depicted in figure 15.4. When a particular locus duplicates itself, two genes with identical base sequences coexist in the same individual. Inasmuch as only one locus is necessary to produce the polypeptide chain, an additional locus might be regarded as superfluous. On the contrary, the duplicated locus might prove valuable in protecting the individual against the hazards of mutational changes. Ordinarily a mutation that interferes adversely with the functioning of a locus would be detrimental or lethal. However, an otherwise lethal mutation could be accepted if it occurred in the duplicated locus, since the original gene is unimpaired and would continue to synthesize the essential polypeptide. With the continual occurrence of mutations, the duplicated locus would produce a polypeptide whose sequence of amino acids is different from the original polypeptide. This new

polypeptide, although still related to the original one, would be sufficiently altered to have a different activity and specificity. In this manner, two polypeptide chains, each with its own gene, would ultimately evolve from a single polypeptide and a single gene.

A particular gene may be duplicated more than once, and large families of related proteins may eventually evolve from a common progenitor. Indeed, it is generally accepted that the varied hemoglobin genes, *α, β, γ,* and *δ,* arose by duplication from a single common ancestral gene. Each duplication was followed by the accumulation of mutations in the duplicated locus. The evolution of different molecular species of a protein thus involved the interplay of gene duplication and mutation. Duplications that originated early in the evolutionary history were frequently separated into different chromosomes by translocations, whereas recently duplicated genes have remained linked on the same chromosome. The *α* and *β* genes of hemoglobin presently reside in separate chromosomes, whereas the more recently evolved *δ* gene has remained closely adjacent to the *β* gene.

EVOLUTIONARY HISTORY OF HEMOGLOBIN

We have thus far suggested that the genes involved in the syntheses of the polypeptide chains of hemoglobin were derived through duplications of a common ancestral locus, followed by divergent mutations. The original hemoglobin molecule probably resembled the chemically related myoglobin molecule. Myoglobin does not function as a respiratory protein; it is used mainly for the storage of oxygen in metabolically active tissues, especially within muscles. The storage of oxygen is very important in diving animals, which have exceptionally high concentrations of myoglobin in their muscles. Mammalian myoglobin is a single polypeptide chain of 153 amino acid residues.

In the distant past, it is likely that myoglobin in tissue cells and hemoglobin in blood cells were identical. Myoglobin evolved into a storage protein

Selective advantage of gene duplication

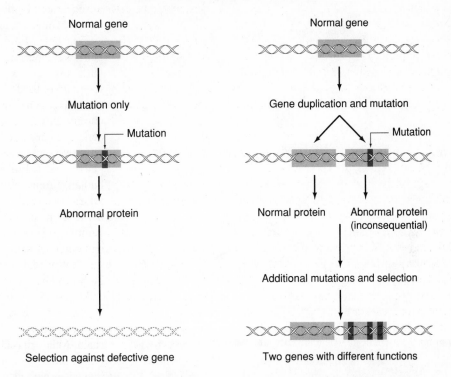

Figure 15.4 Duplication of a gene followed by random mutations and natural selection can lead to two genes with different functions. In the absence of gene duplication, a mutation would be lethal if the mutant gene produces an aberrant protein *(left).* With a duplicated gene, the same mutation that would otherwise be lethal is not selectively disadvantageous, since the original gene continues to produce the normal protein *(right).* With time, several mutations can accumulate in the duplicated gene.

with a high oxygen affinity, and hemoglobin became progressively adapted to fulfill its primary role of oxygen transport in different species. Based on the degree of similarity (or homology) of the amino acid residues between any two polypeptide chains, an evolutionary tree can be constructed (fig. 15.5). Each branching point in the tree indicates where an ancestral locus was duplicated, giving rise each time to a new gene line.

Linus Pauling calculated that one amino acid substitution has occurred every 7 million years. The numerous differences in the amino acid sequences between myoglobin and hemoglobin indicate that the two molecules had a common ori-

gin at least 650 million years ago, well before the first appearance of the vertebrates. A one-chain hemoglobin still persists in primitive fish, the cyclostomes. While the cyclostomes (lamprey and hagfishes) have a single hemoglobin chain, the carp (bony fish) has two chains. Accordingly, the differentiation of the alpha and beta chains of hemoglobin occurred about 500 million years ago, when the first bony fishes appeared. At that time, the alpha and beta chains became aggregated to produce the efficient four-chain hemoglobin molecule characteristic of most vertebrates.

During mammalian evolution, some 200 million years ago, the β locus duplicated once again

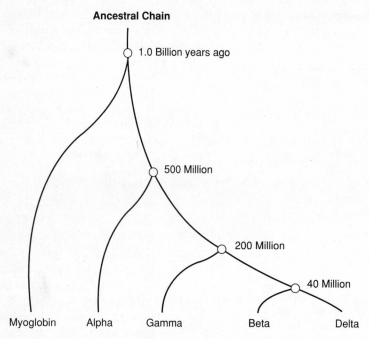

Ancestral Chain

1.0 Billion years ago

500 Million

200 Million

40 Million

Myoglobin   Alpha   Gamma   Beta   Delta

Figure 15.5 Molecular evolution of hemoglobin chains. The small circles at the foci of branching indicate where the ancestral genes were duplicated to give rise to new globin chains.

to give rise to the γ gene that codes for fetal hemoglobin with its additional oxygen-binding capacity. The most recent duplication resulting in the δ gene occurred only 40 million years ago. The delta chain has been found only in higher primates. The divergence of the δ gene occurred prior to the separation of the phylogenetic lineage leading to the Old World monkey assemblage and the line represented by Old World monkeys, great apes, and humans. Curiously, present-day Old World monkeys do not synthesize delta chains, at least in detectable amounts. Apparently, the δ gene became silenced recently in Old World monkeys.

PSEUDOGENES

Genetic studies have established that the α gene and the β gene reside in different chromosomes. Indeed, a cluster of α-like genes has evolved from a single ancestral α gene by a series of duplications in one chromosome, and a comparable family of β genes has emerged in another chromosome. As shown in figure 15.6, the linked group of α-like genes on human chromosome 16 contains two active embryonic zeta (ζ) and two active alpha (α) genes. The β-like cluster on human chromosome 11 comprises five active genes—namely, one epsilon (ε), two gamma (γ), one delta (δ), and one beta (β).

Both families of hemoglobin genes contain loci called *pseudogenes* (depicted as *psi, ψ*) that cannot encode a functional polypeptide chain. Although each pseudogene shares many base sequences with its corresponding normal gene, the presence of a so-called "frameshift" mutation has shifted the triplets of bases so that polypeptide synthesis is prematurely halted. Pseudogenes apparently are products of gene duplication that have accumulated debilitating base changes during sequence divergence. If the $\psi\alpha1$ gene were to become functional in the future, then the alpha chain would be coded by three operational α genes.

The occurrence of two immediately adjacent genes, both of which are expressed, is a common feature of gene clusters. Thus, two α genes ($\alpha1$ and $\alpha2$) are expressed in the production of alpha polypeptides. Such duplication of virtually identical genes has proved to be providential to some patients afflicted with the blood disorder, *thalassemia,* characterized by a reduced rate of alpha chain synthesis as a consequence of a defective α gene. Some synthesis of alpha chains becomes possible if the impairment involves only one of the several α loci.

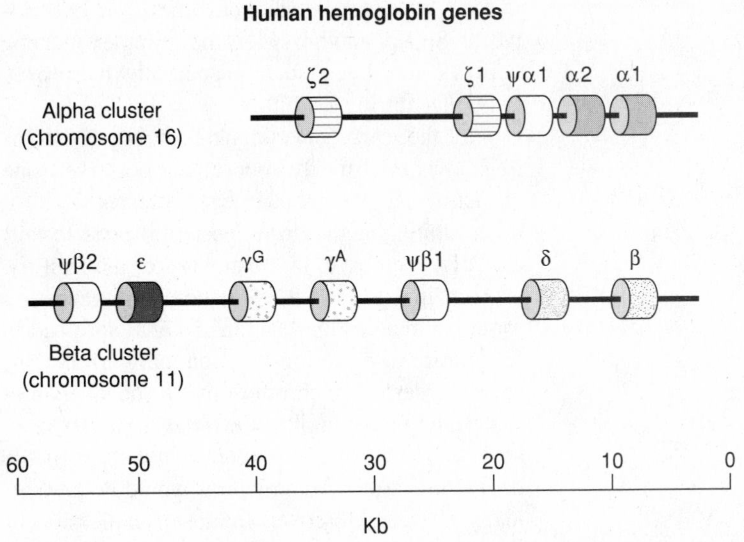

Human hemoglobin genes

Figure 15.6   The two families of human hemoglobin genes form clusters on different chromosomes. The *alpha* cluster of linked genes occupies a length of about 30 Kilobases (30,000 base pairs). The *beta* cluster of genes is spread over a region of about 50 Kb (50,000 base pairs).

The genes in both hemoglobin linkage groups are separated by substantial tracts of intergenic DNA. In other words, only about 10 percent of the DNA in these clusters actually code for hemoglobin polypeptides. The function of the remaining 90 percent of intergenic DNA between the genes is presently enigmatic. Some investigators have suggested that these extensive intergenic regions might have no function and actually represent "junk" DNA. If such DNA is meaningless, then the base sequences could, during evolution, change rapidly without consequences. In this vein, it is of interest that the arrangements of the entire β cluster of genes are indistinguishable in the gorilla, baboon, and humans. This indicates that the arrangement of the human cluster has been fully established 20 to 40 million years ago and has been faithfully preserved since. This conservation of the arrangement suggests that the base sequences throughout the β region have been substantially constrained by selection, at least during recent primate evolution. Such conservative evolution is incompatible with the notion that intergenic DNA is "junk." One of the pioneers of molec-

ular biology, Sidney Brenner, has cogently reminded us of a valuable difference between "junk" and "garbage"— garbage is thrown away whereas junk is stored (in the garage) for some unforeseen future use.

HOMEOTIC GENES

The name of Edward B. Lewis of the California Institute of Technology immediately brings to mind his pioneering studies more than 50 years ago on genetic changes in the fruit fly, changes known as *homeotic* mutations. These mutations are responsible for such dramatic developmental effects as the transformation of parts of the body into structures appropriate to other positions. Legs may form in a position normally bearing a wing, or a second set of wings may emerge in addition to the usual single pair of wings (fig. 15.7). Some evolutionists viewed such drastic morphological changes as major mutations, or *macromutations*. In the 1940s, the prominent geneticist Richard Goldschmidt contended that homeotic mutants illustrate a possible mode of origin of novel types of body structure and hence could be responsible for the rapid emergence of a new species in the distant past. Goldschmidt's views are considered untenable today, but Lewis's superb findings remain a landmark in developmental genetics.

The inevitable question is: What is the normal role of a homeotic gene? The recent revolution in molecular biology has revealed that homeotic genes do have assignments that are the antithesis of the execution of bizarre disruptions of structures. Homeotic genes, in their normal state, are distinctly "master" genes that orchestrate the orderly unfolding of the genetic program for the

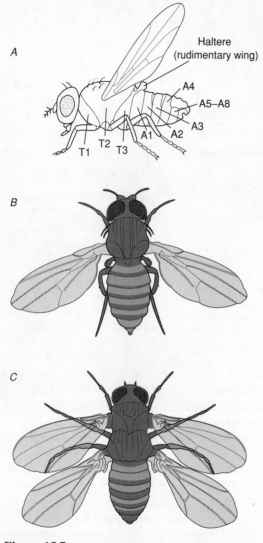

Figure 15.7 *(A)* A drawing of a normal fly. The third thoracic segment *(T3)* has a rudimentary wing, known as a *haltere*. *(B)* A normal fly with a single set of wings. *(C)* Homeotic mutations transform segment *T3* into a structure similar to *T2*. A second, fully developed set of wings emerges from this transformed segment.
(Based on studies of E. B. Lewis, 1978, *Nature* 276:565.)

development of the embryo. Perhaps unthinkable, these master developmental genes are shared in animals ranging from the fruit fly to humans, which suggests that present-day life forms utilize

in their development an ancient basic genetic network. Stated another way, the complex genetic network for development apparently has arisen only once during evolution.

In the early development of organisms as diverse as the fruit fly, mouse, and ourselves, one particular set of homeotic genes is responsible for determining the anterior to posterior (head-to-tail) axis of the embryo. In the embryo of the fruit fly, as seen in figure 15.8, eight homeotic genes are arranged in a tandem fashion and are expressed in an anterior to posterior direction that mirrors their spatial order in the chromosome. If the most anterior gene of the complex, the *labial* gene (abbreviated "lab" in fig. 15.8) were to mutate, then the development of head structures would be dramatically altered. Moreover, failure of expression of the *labial* gene by virtue of a mutation could lead to an overexpression of adjacent genes in the complex, resulting in strange imperfections in more posterior parts of the body. Certain homeotic genes normally act by suppressing the development of structures. One of the homeotic genes expressed in the posterior thorax inhibits the development of a posterior (second) wing. Consequently, *halteres* (balancing organs resembling a reduced wing) are produced at that site rather than a second pair of wings.

The mechanism by which a homeotic gene exerts its effects has been actively pursued. It is now known that the protein product of a homeotic gene acts as a *transcription factor*. A transcription factor allows the cell to differentially turn on the expression of a particular target gene. It does so by binding to specific DNA sequences adjacent to the target gene it regulates. In essence, homeotic genes are master switches that bring into play, and coordinate the action of, other developmental genes. Each homeotic gene serves to engender a cascade of developmental effects.

Genes homologous to the homeotic genes of the fruit fly have been demonstrated in the mouse and humans. The normal homeotic genes of the fruit fly are collectively known as the homeotic complex, or HOM-C. The homologous genes in vertebrates are said to be *Hox* genes. Irrespective

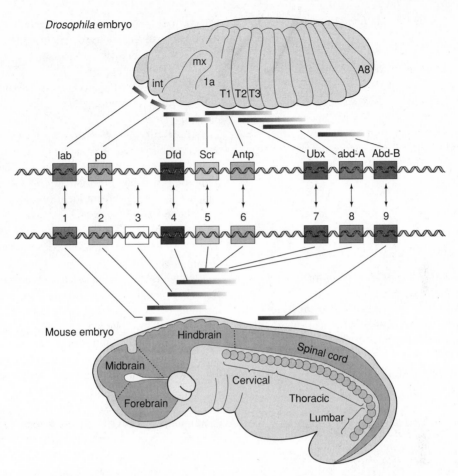

Figure 15.8 **Similar spatial organization of homeotic genes in the fruit fly and mouse embryos.** Segmental expression of the cluster of *Hox* genes in the mouse embryo is most pronounced in the developing central nervous system (nerve cord) and developing bony spinal column.

(From John Gerhart and Marc Kirschner, *Cells, Embryos, and Evolution,* Blackwell Science, 1997, fig. 7.12.)

of nomenclature, the important message is that all HOM-C and *Hox* genes share a common ancestry. There exist in nature sets of structurally similar genes, originally discovered in the fruit fly, that are the prime regulators of embryonic development in life forms.

During the 1980s, homeotic genes were isolated and cloned in several species. A specific nucleotide sequence, called the *homeobox* sequence, is common to the coding region of all HOM-C and *Hox* genes. This sequence, 180

nucleotides in length, can be used as a genetic probe. The homeobox probe cross-hybridizes with the genomic DNA of many organisms, including yeast and humans. The homeobox sequences of the fruit fly show 90 percent homology with the homeobox sequences discovered in human genes. The conserved homeobox sequence apparently is a general feature of many developmental regulatory genes.

The homeobox sequence encodes a protein segment with 60 amino acid residues. The protein segment is known as the *homeodomain,* the part of the protein that binds strongly to recognition sites in other genes. Approximately 30 percent of the amino acids in the homeodomain are basic— arginine and lysine. This is a hallmark of DNA-binding regions of proteins. As seen in table 15.2, the homeodomains of various species are so highly conserved that they differ only in tolerable amino acid substitutions—that is, the replacement of one amino acid by another of similar chemical properties.

OCULAR MALFORMATIONS AND EVOLUTION

There is another family of developmental control genes, called *Pax* genes, that encode transcription factors. The vertebrate *Pax* genes are key regulators of the development of various organs of the body, notably the eye. One particular gene member of the family, *Pax-6,* has earned notoriety for causing malformations of the eye in mice

and humans. Mutations in the *Pax-6* gene are responsible in the mouse for diminutive eyes in the heterozygous state and the complete lack of eyes in the homozygous condition. The *Pax-6* homologue in the human has been shown to be mutated in patients with congenital *aniridia* (lack of iris).

The *Pax-6* gene is characterized by a highly conserved homeobox-like sequence, a span of 130 amino acids coded by 390 nucleotides. The homologue of *Pax-6* has been isolated in the fruit fly, which shares a 94 percent identity with the amino acid sequences of mice and humans. Not unexpectedly, the *Pax-6* homologue maps to the *eyeless (eye)* locus on the fourth chromosome of *Drosophila.*

Notwithstanding the great anatomical differences between the eyes of insects and vertebrates, the mouse *Pax-6* gene induces ectopic eye development when transplanted into fruit fly embryos. It would seem that the fundamental role of the *Pax-6* gene in eye development has been conserved throughout the animal kingdom. After hundreds of

TABLE 15.2	Extensive Homology in Amino Acid Sequences of Five Homeodomains

	1																			20
Mouse (MO-010)	Ser	Lys	Arg	Gly	Arg	Thr	Ala	Tyr	Thr	Arg	Pro	Gln	Leu	Val	Glu	Leu	Glu	Lys	Glu	Phe
Frog (MM3)	Arg	Lys	Arg	Gly	Arg	Gln	Thr	Tyr	Thr	Arg	Tyr	Gln	Thr	Leu	Glu	Leu	Glu	Lys	Glu	Phe
Fruit fly (Antennapedia)	Arg	Lys	Arg	Gly	Arg	Gln	Thr	Tyr	Thr	Arg	Tyr	Gln	Thr	Leu	Glu	Leu	Glu	Lys	Glu	Phe
Fruit fly (Fushi tarazu)	Ser	Lys	Arg	Thr	Arg	Gln	Thr	Tyr	Thr	Arg	Tyr	Gln	Thr	Leu	Glu	Leu	Glu	Lys	Glu	Phe
Fruit fly (Ultrabithorax)	Arg	Arg	Arg	Gly	Arg	Gln	Thr	Tyr	Thr	Arg	Tyr	Gln	Thr	Leu	Glu	Leu	Glu	Lys	Glu	Phe

	21																			40
Mouse (MO-10)	His	Phe	Asn	Arg	Tyr	Leu	Met	Arg	Pro	Arg	Arg	Val	Glu	Met	Ala	Asn	Leu	Leu	Asn	Leu
Frog (MM3)	His	Phe	Asn	Arg	Tyr	Leu	Thr	Arg	Arg	Arg	Arg	Ile	Glu	Ile	Ala	His	Val	Leu	Cys	Leu
Fruit fly (Antennapedia)	His	Phe	Asn	Arg	Tyr	Leu	Thr	Arg	Arg	Arg	Arg	Ile	Glu	Ile	Ala	His	Ala	Leu	Cys	Leu
Fruit fly (Fushi tarazu)	His	Phe	Asn	Arg	Tyr	Ile	Thr	Arg	Arg	Arg	Arg	Ile	Asp	Ile	Ala	Asn	Ala	Leu	Ser	Leu
Fruit fly (Ultrabithorax)	His	Thr	Asn	His	Tyr	Leu	Thr	Arg	Arg	Arg	Arg	Ile	Glu	Met	Ala	Tyr	Ala	Leu	Cys	Leu

	41																			60
Mouse (MO-10)	Thr	Glu	Arg	Gln	Ile	Lys	Ile	Trp	Phe	Gln	Asn	Arg	Arg	Met	Lys	Tyr	Lys	Lys	Asp	Gln
Frog (MM3)	Thr	Glu	Arg	Gln	Ile	Lys	Ile	Trp	Phe	Gln	Asn	Arg	Arg	Met	Lys	Trp	Lys	Lys	Glu	Asn
Fruit fly (Antennapedia)	Thr	Glu	Arg	Gln	Ile	Lys	Ile	Trp	Phe	Gln	Asn	Arg	Arg	Met	Lys	Trp	Lys	Lys	Glu	Asn
Fruit fly (Fushi tarazu)	Ser	Glu	Arg	Gln	Ile	Lys	Ile	Trp	Phe	Gln	Asn	Arg	Arg	Met	Lys	Ser	Lys	Lys	Asp	Arg
Fruit fly (Ultrabithorax)	Thr	Glu	Arg	Gln	Ile	Lys	Ile	Trp	Phe	Gln	Asn	Arg	Arg	Met	Lys	Leu	Lys	Lys	Glu	Ile

From W. J. Gehring. 1985. *Sci. Am.* 253/4:159.

millions of years of evolutionary independence of the fruit fly and mouse, the astounding similarity between the developmental genes of an insect and a mammal seems possible *only* if these major transcriptional genes arose *only once* in a distant common ancestor.

The time-honored view is that the compound eye of insects and the camera-like eye of vertebrates evolved independently. Historically, evolutionary biologists have interpreted the origin of eyes in well-known lineages—squid, insects, and vertebrates—as illuminating examples of *convergence,* or the evolution of similar traits in genetically unrelated species, chiefly as a result of similar environmental selective pressures. In 1994 at the University of Basel in Switzerland, a team of developmental biologists led by Walter J. Gehring, proposed that the appearance of eyes in varied groups reflects *shared ancestry* (homologies of developmental genetic networks) and not *independent evolution* (convergence). This revisionistic paradigm has been embraced by some outstanding evolutionists, notably Stephen J. Gould of Harvard University, but has been challenged by other prominent evolutionists, particularly Ernst Mayr, emeritus professor at Harvard University. Mayr contends that photoreceptive organs have evolved independently on innumerable occasions, perhaps 60 or more times. He suspects that the primitive genetic network has been continually modified through time and that various genetic components of the conserved pathways have been extensively recruited in new combinations for other adaptive purposes in different lineages.

MULTILEGGED FROGS REVISITED

We have seen that embryonic development is governed by a repertoire of transcription factors coded by master genes. To add an additional facet to the story, there are signals from the outside of developing cells that can modulate the activity of the transcription factors. These signals take the form of *signaling macromolecules* that enter cells and profoundly influence the differentiation of cells,

tissues, and organs. For example, normal limb development in vertebrates depends upon a threshold concentration of endogenous retinoic acid (an analog of vitamin A) reaching the developing limb bud at the precise time that *Hox* genes are being transcribed.

In 1993, Malcolm Maden of King's College in London reported the striking homeotic transformation of tails into hindlimbs of frog tadpoles treated with retinoic acid. At a certain stage of development, the tails of tadpoles of a European species of frogs *(Rana temporaria)* were amputated and the experimental animals were placed for several days in varied concentrations of retinoic acid. At a particular dose and treatment time, the tail actually

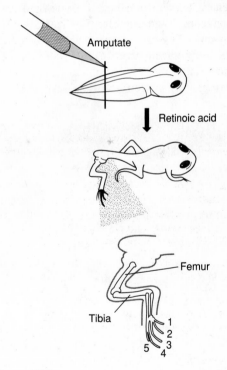

Figure 15.9 The tip is amputated from a tadpole's tail and the stump is exposed to retinoic acid. The tail stump regenerates a hindlimb rather than a tail. The anatomy of the hindlimb is normal; the pictorial enlargement of the limb shows the normal five digits. (Based on studies by M. Maden.)

regenerated a normal hindlimb (fig. 15.9). Control animals—with tails amputated but reared in the absence of retinoic acid—regenerated normal tails.

Retinoic acid may be viewed as a developmental teratogen in the sense that elevated levels can alter, or redirect, the developmental pathways. In our introductory first chapter, we evoked the usual environmental suspects—pesticides, for example—to account for the multilegged anomaly in frogs collected in nature. Now we invite the reader to consider the possibility that the causative factor in nature may have been the sensitivity of *Hox* genes to inappropriate levels of retinoic acid (or some other *Hox*-sensitive substance). Parenthetically, it may be noted that retinoic acid is occasionally used as a medication to treat skin disorders in humans, disorders such as acne. Pregnant women have been cautioned to avoid taking retinoic acid in pill form. When absorbed in large quantities by a developing fetus, retinoic acid is highly teratogenic, causing both limb and cardiac malformations.

"MITOCHONDRIAL EVE"

Previously, in chapter 14, we stated that mitochondria probably began their existence as free-living, aerobic bacteria that had been engulfed by a primitive eukaryote. Mitochondria have a number of unique features in modern eukaryotic cells. For the present discussion, we shall cite two of their prominent features. First, the mutation rate of mitochondrial DNA is high, primarily because mitochondria have no effective DNA repair mechanism. The high mutation rate may also reflect the large concentration of mutagens (such as superoxide radicals) resulting from the intense metabolic functions performed by mitochondria. Second, mitochondrial DNA is inherited exclusively from the mother, and thus complications due to allelic segregation (as with nuclear genes) do not exist. The unfertilized egg contains some 200,000 maternal mitochondria, whereas a sperm cell contains a relatively small number of mitochondria

localized in the middle piece. The sperm's middle piece disintegrates during the act of fertilization, so that the father cannot contribute to the ultimate mitochondrial genotype of the offspring.

Maternal inheritance coupled with the high mutation rate of mitochondrial DNA make it possible to distinguish among closely related individuals and, in fact, to trace human origins. All present-day humans may have shared a common ancestral mother (our "mitochondrial Eve"). One can attempt to reconstruct human female phylogeny by detecting and quantifying nucleotide mutational changes by means of restriction enzymes. There are 37 genes, encompassing 16,569 pair bases in each circle of mitochondrial DNA.

In the 1980s, Rebecca L. Cann and Mark Stoneking, working under the direction of Allan C. Wilson at the University of California, analyzed mitochondrial DNA variations in women representing five geographical groups: Africa, Asia, Europe (Caucasian), Australia (aboriginal Australians), and New Guinea (aboriginal New Guineans). The cardinal result from the genetic samples of women around the world is that sub-Saharan African populations exhibit *greater genetic diversity* than non-Africans. Populations other than those in Africa retain just a subset of this genetic diversity, as would be expected when a smaller group breaks away from a founding population, taking only a small sample of the full range of genetic variation.

The ancestral mitochondrial sequence apparently arose in Africa (fig. 15.10). Stated another way, the number of intercontinental migrations required to account for the present-day geographic distribution of mitochondrial DNA types is best explained by postulating that the source of the human mitochondrial gene pool is Africa. This is in accord with the thesis, subscribed to by many evolutionists, that humans first emerged in Africa and then emigrated to other continents. The mitochondrial Eve is presumed to have lived in Africa about 150,000 years ago. It is likely that there has been more than one migration out of Africa. This theme will be explored in chapter 17.

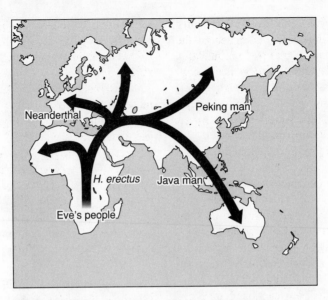

Figure 15.10 The migration of Eve's people, a group of modern humans who emerged in sub-Saharan Africa 150,000 years ago and then swept across Europe, India, Asia, and Australia. Presumably, the invading Eve's people replaced previously existing human groups, such as the Neanderthals in Europe, Peking man in Asia, and Java man in Australia and Indonesia (see chapter 17).

(Illustration from James Shreeve, "Argument Over a Woman," *Discover* [August 1996]:52–59.)

Data on mitochondrial DNA have enabled investigators, among them Douglas C. Wallace of Emory University, to reconstruct the occupation of the Americas by northeast Asian migrants. Clearly, American aboriginal peoples did not evolve *de novo* on the American continents. The Americas were populated by only a few migrant groups from Asia, and the first of these migrations probably occurred 40,000 years ago or less. The most ancient migrants are represented today by descendants who speak "Amerind" languages. The Amerinds, who comprise most native Americans, entered the Americas in a single migratory wave 20,000 to 40,000 years ago. The Navajo, Apache, and other members of a Native American group known collectively as the "Na-Dene," are latecomers. The ancestors of modern speakers of Na-Dene entered the continent in a second migration, a mere 5,000 to 10,000 years ago.

16

HISTORY OF LIFE

Our interpretation of the ever-changing procession of life is based in great measure on the fossils of animals and plants entombed in the earth's rocks. An organism becomes preserved as a fossil when it is trapped in soft sediment that settles at the bottom of a lake or ocean. The deposits of mud and sand harden into the sedimentary, or stratified, rocks of the earth's crust. Fossils, then, are the remains of past organisms preserved or imprinted in sedimentary rocks. Not all fossils are mere impressions of buried parts. Exceptionally, the whole organism may be preserved, as attested by the woolly mammoths embedded in the frozen gravels of Siberia. In the case of petrified tree trunks, the wood was infiltrated with mineral substances that crystallized and hardened. Incredibly, such mineral-impregnated wood can be sliced in very thin sections and studied microscopically for details of cellular structures.

The fossil remains of past life are incomplete and uneven. Innumerable organisms with soft tissue (jellyfishes and flatworms, for example) are not good candidates for preservation. In contrast, the woody parts of plants, the shells of mollusks, and the bones and teeth of vertebrates are relative-

ly common as fossils. In many places of the earth, the sedimentary rocks have been subjected to such pressure and heat that their fossil holdings have been irretrievably lost. The earth's rocks have also been exposed to the relentless forces of weathering and erosion. Despite the destructive work of geologic processes, the available fossil assemblage contains an extraordinary amount of information. The older strata of rock, those deposited first, bear only relatively simple kinds of life, whereas the newer, or younger, beds contain progressively more and more complex types of life. The increasing complexity of fossils in progressively more recent rock strata is one of the more profound evidences for evolutionary changes in organisms with time. Each species of organisms now living on the earth has emerged from an earlier ancestral form.

GEOLOGIC AGES

The arrangement of geologic time is recorded in table 16.1. The main divisions, or *eras,* embrace a number of subdivisions, or *periods,* and these periods are further refined into *epochs.* Fossils appear in abundance in the rocks of the Cambrian period,

TABLE 16.1	Geologic Time Scale				
Era	**Period**	**Epoch**	**Millions of Years Ago**	**Important Events**	
	Quaternary	Recent	0.01	Modern humans	
		Pleistocene	2	Early humans	
Cenozoic	Tertiary	Pliocene	10	Radiation of apes	
(Age of Mammals)		Miocene	25	Abundant grazing mammals	
		Oligocene	35	Angiosperms dominant	
		Eocene	55	Mammalian radiation	
		Paleocene	75	First placental mammals	
Mesozoic	Cretaceous		135	Climax of reptiles, first angiosperms, extinction of ammonoids	
(Age of Reptiles)	Jurassic		185	Reptiles dominant, first birds, first mammals	
	Triassic		225	First dinosaurs, cycads and conifers dominant	
	Permian		275	Widespread extinction of marine invertebrates, expansion of primitive reptiles	
Paleozoic	Carboniferous*		350	Great swamp trees (coal forests), amphibians prominent	
	Devonian		400	Age of fishes, first amphibians	
	Silurian		430	First land plants, eurypterids prominent	
	Ordovician		480	Earliest known fishes	
	Cambrian		600	Abundant marine invertebrates, trilobites and brachiopods dominant, algae prominent	
Precambrian			>3000	Soft-bodied primitive life	

*The early Carboniferous is often referred to as the "Mississippian" and the late Carboniferous as the "Pennsylvanian."

which dawned 600 million years ago. The Cambrian strata bear numerous fossils of marine plants and animals; seaweeds, sponges, shellfishes, starfishes, and bizarre arthropods called trilobites. One of the more interesting Cambrian fossil sites is the *Burgess Shale,* located in British Columbia in Canada. Dated about 530 million years ago, the Burgess Shale includes an impressive assemblage of hard-bodied and soft-bodied creatures. The soft-bodied Burgess fossils owe their preservation to quick burial under anaerobic (non-oxygen) conditions.

The Cambrian was a period of great geologic upheaval of the earth's surface. Mountains were being formed as continents collided. Earthquakes and ice ages added to the flux of environmental conditions. Oxygen levels in the oceans and the atmosphere were on the rise. Whether these earth-

ly upheavals fostered the burst and diversification of life is unknown. Yet, it is likely that the Cambrian infusion of diversity was a function of the vast openness of ecological niches. The most explosive period of Cambrian life, estimated between 530 and 520 million years ago, has been aptly described as "biology's Big Bang."

Cambrian life did not come into existence abruptly. The evolutionary epic began at a far more distant time. At least 3 billion years of slow organic evolution preceded the diversity of organisms of the Cambrian period. The Precambrian seas swarmed with a great variety of microscopic forms of life. Studies in recent years have dispelled the notion that Precambrian rocks are devoid of fossils. Precambrian life was dominated by prokaryotic bacteria and blue-green algae (cyanobacteria). The oldest microfossil life has been uncovered from Warrawoona in Western Australia and dates back 3.6 billion years ago. The bacteria-like microfossils of Warrawoona are remarkedly similar to present-day blue-green algae. In addition, the electron microscope has been used to examine samples of deep sedimentary rock from South Africa (the so-called Fig Tree sediments). The microscopic analysis uncovered impressions of rod-shaped bacteria that existed 3.2 billion years ago. Further, minute traces of two chemical substances, phytane and pristane, have been found in the Fig Tree rocks. These chemicals are the relatively stable breakdown products of chlorophyll. Thus, photosynthesis by algae, in which oxygen is released, may have begun about 3 billion years ago. Other ancient rocks—such as the 2-billion-year-old Gunflint rock formation in Ontario, Canada—contain an assemblage of fossil microorganisms that appear to be blue-green algae, possibly red algae, and even some fungi. Eukaryotic organization is clearly in evidence in deposits that are approximately one billion years old. The Bitter Springs formation of Australia contains eukaryotic fossils that represent higher algae and fungi. The transition from prokaryotes to eukaryotes apparently occurred during the late Precambrian era.

We may visualize the vast span of time by compressing the history of life into 60 minutes. In the 60-minute clock depicted in figure 16.1, each second represents one million years. The Precambrian era is conservatively viewed as originating at 12 o'clock, or 3.6 billion years ago. Primitive bacteria have left imprints in the sediments 3.2 billion years ago, or about 7 minutes after the start of the clock. The Precambrian era stretches over five-sixths of the earth's entire history, during which time primitive life was confined to the sea.

Toward the close of the Precambrian 700 million years ago (or 11 minutes and 40 seconds before the hour), the fossil-bearing rocks show traces of soft-bodied invertebrates of simple organization. At the onset of the Paleozoic era 600 million years ago, the fossil record improves immeasurably. The Paleozoic era begins with only 10 minutes left in the hour and occupies a span of only 6 minutes and 15 seconds (or 375 million years). Nevertheless, within this short interval, we witness the invasion of land by plants, the emergence of great swamp trees, and the dominance of fishes and amphibians.

The Mesozoic era, during which the reptiles reigned, commences with less than 4 minutes left to the hour and lasts for 2 minutes and 30 seconds (150 million years). Mammals and flowering plants dominate the landscape of the Cenozoic era during the last 1 minute and 15 seconds of the clock hour. Not until the closing seconds of the evolutionary hour does the human species appear on the scene. The whole span of recorded human history covers the last 3 seconds!

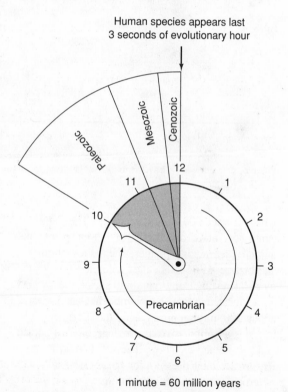

1 minute = 60 million years

Figure 16.1 Model of evolutionary time, where one second equals one million years. The Precambrian is represented as originating at 12 o'clock, or 3.6 billion years ago. The human species appeared about 3 million years ago, or a mere terminal three seconds of the evolutionary hour.

FOSSIL RECORD OF PLANTS

The general character of plant life from the dawn of the Cambrian period to recent times is depicted in figure 16.2. Only the broader aspects of past evolution are portrayed. It is evident that the flora has changed in composition over geologic time. New types of plants have continually appeared. Some types thrived for a certain time and then disappeared. Others arose and flourished. Still others have survived until the present, only in much reduced numbers.

The Cambrian and Ordovician periods were characterized by a diverse group of algae in the warm oceans and inland seas. One of the most significant advances, which occurred during the Silurian period, was the transition from aquatic existence to life on land. The first plants that established themselves on land were diminutive herbaceous forms, the *psilophytes,* literally "naked plants," in allusion to their bare, leafless stems. The primitive woody stems sent up swollen aerial shoots to scatter their spores and thrust branches down into the earth for water. The existence of psilophytes made possible the subsequent emergence of an infinite variety of tree-sized plants that flourished in the swamps of Carboniferous times. Two prominent trees that luxuriated in Carboniferous forests were the giant club moss, *Lepidodendron,* which soared to heights of 130 feet on slender trunks, and the coarse-leaved *Cordaites,* progenitors of the modern conifers. The

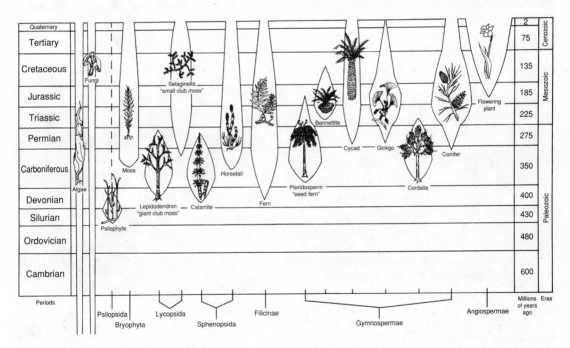

Figure 16.2 Historical record of plant life. Plant life during the Cambrian and Ordovician, the first two periods of the Paleozoic era, was confined to the water; seaweeds (algae) of immense size, often several hundred feet in length, dominated the seas. Land plants came into existence in Silurian time, in the form of strange little vascular plants, the psilophytes. In the Carboniferous period, imposing spore-bearing trees (Lepidodendrids and Calamites) and primitive naked-seeded plants (Pteridosperms and Cordaites) reached their peak of developement. The end of the Paleozoic era marked the extinction of the majority of the luxuriant trees of the Carboniferous coal swamps. The Mesozoic era was the *Age of Gymnosperms,* as evidenced by the abundance of cycads, ginkgoes, and conifers. Flowering plants (Angiosperms) rose to ascendancy toward the close of the Mesozoic era and established themselves as the dominant plant group on the earth.

first seed-bearing plants, *Pteridosperms,* arose in the humid environment of the vast swamps. These ancient groups of tall trees dwindled toward the close of the Paleozoic era and shortly became extinct. Their decomposed remains led to the formation of extensive coal beds throughout the world. Figure 16.3 shows the appearance of a Carboniferous swamp as reconstructed from fossils.

By far the greater number of Paleozoic species of plants failed to survive. The Paleozoic flora was largely replaced by the seed-forming gymnosperms, which became prominent in the early Mesozoic era. A diverse assemblage of cycads, ginkgoes (maidenhair trees), and conifers formed elaborate forests. During the closing years of the Mesozoic, the flowering plants (angiosperms), which began very modestly in the Jurassic period, underwent a phenomenal development and today constitute the dominant plants of the earth. The

angiosperms have radiated into a variety of habitats, from sea level to mountain summits and from the humid tropics to the dry deserts. Associated with this diversity of habitat is a great variety of general form and manner of growth. Many angiosperms have reverted to an aquatic existence. The familiar duckweed, which covers the surface of a pond, is a striking example.

FOSSIL RECORD OF ANIMALS

In the animal kingdom, we witness a comparable picture of endless change, with waves of extinction and replacement (fig.16.4). The deep Cambrian rocks contain the remains of varied invertebrate animals—sponges, jellyfishes, worms, shellfishes, starfishes, and crustaceans. One of the major developments of the Cambrian period was the advent of hard parts in the form of protective shells

Figure 16.3 Luxuriant forests of giant trees with dense undergrowth flourished in the Great Coal Age (Carboniferous period), between 350 million and 275 million years ago. The massive trunks at the left are the Lepidodendrids, an extinct group whose modern relatives include the small, undistinguished club mosses *(Selaginella)* and the ground pines *(Lycopodium)*. The tall, slender tree with whorled leaves at the right is a Calamite, represented today by a less prominent descendant, the horsetail *Equisetum*. The fernlike plants bearing seeds (at the left) are seed ferns (Pteridosperms), which resembled ferns but were actually the first true seed-bearing plants (Gymnosperms).

(Courtesy of Field Museum of Natural History, Chicago.)

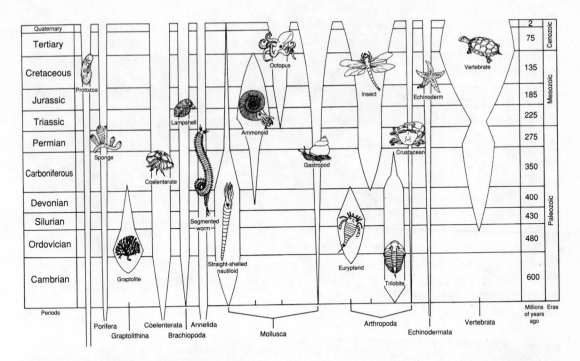

Figure 16.4 Historical record of animal life. Many of the invertebrate groups were already highly diversified and abundant in the Cambrian, the first period of the Paleozoic era, approximately 600 million years ago. The Paleozoic era is often called the *Age of Invertebrates*, with its multitude of nautiloids, eurypterids, and trilobites. Brachiopods with hinged valves were the commonest shellfish of the Paleozoic seas. In the Mesozic era, air-breathing insects and vertebrates, notably the widely distributed reptiles, held the center of the stage. The Mesozoic seas were populated with large, shelled ammonoids, now extinct. Warm-blooded vertebrates (birds and mammals) became prominent in the Cenozoic era, and the human species arrived on the scene in the closing stages of this era.

and plates. The shelled invertebrates were already so well developed that their differentiation must have taken place during the long period preceding the Cambrian. That this is actually the case has been attested by the discoveries of rich deposits of Precambrian fossils in rocks in the Ediacara Hills in southern Australia and on Victoria Island in the Canadian Arctic. These fossil-bearing rocks lie well below the oldest Cambrian strata and contain imprints of jellyfishes, tracks of various kinds of worms, and traces of soft corals. Since the rocks on Victoria Island are 700 million years old, these discoveries extend the fossil record of invertebrates 100 million years beyond the Cambrian.

Dominating the scene in early Paleozoic waters were the crablike trilobites and the large scorpionlike eurypterids (fig.16.4). Most trilobites were not more than 3 inches long; the so-called giant trilobite did not exceed 18 inches. The most powerful dynasts of the Silurian waters were the eurypterids, which occasionally attained a length of 12 feet. Common in all Paleozoic periods were the nautiloids, related to the modern *Nautilus*. The clamlike brachiopods, or lampshells, constitute nearly one-third of all Cambrian fossils, but are relatively inconspicuous today. The odd graptolites, colonial animals whose carbonaceous remains resemble pencil marks, attained the peak of their development in the Ordovician period and then faded away. No land animals are known from Cambrian and Ordovician times. Seascapes of the early Paleozoic are shown in figure 16.5.

Figure 16.5 Highly diversified assemblage of invertebrates of early Paleozoic seas. *Top:* Ordovician period. Large organism in foreground is a straight-shelled nautiloid. Other prominent forms are trilobites, massive corals, smaller nautiloids, and a snail. *Bottom:* Cambrian period. Conspicuous animals are the trilobite (center foreground), eurypterid (center background), and the jellyfish (left). Other animals include brachiopods, annelid worms, sea cucumber, and varied shelled forms.

(*Top,* courtesy of Field Museum of Natural History, Chicago: *bottom,* courtesy of American Museum of Natural History, New York City.)

The bony fishes arose in the Devonian, the Age of Fishes. The lobe-finned fishes, or crossopterygians, had strong muscular fins by which they crawled, in times of drought, from pool to pool along arid streambeds. These air-breathing crossopterygians gave rise to the first land-dwellers among vertebrates, the amphibians. The end of the Paleozoic era witnessed a decline in amphibians and the initiation of reptilian diversification.

Many of the prominent Paleozoic marine invertebrate groups became extinct or declined sharply in numbers before the Mesozoic era. During the Mesozoic, shelled ammonoids thrived in the seas and insects became prominent. Some of the ancient insects were enormous, like *Meganeuron,* a form of dragonfly with a wing-spread of 29 inches. Among vertebrates, the reptiles radiated swiftly to dominate the land. At the close of the Mesozoic, the once successful marine ammonoids perished and the reptilian dynasty collapsed, giving way to the birds and mammals. Insects have continued to thrive and have differentiated into a staggering variety of species. Well over 800,000 different species of insects have been described, and conservative estimates place the total number of living species today at 3 million.

During the course of evolution, plant and animal groups interacted to each other's advantage. There is little doubt, for example, that the rise and spread of flowering plants fostered the diversification and dispersal of insects. As flowering plants became less and less dependent on wind for pollination, a great variety of insects emerged as specialists in transporting pollen. The colors and fragrances of flowers evolved as adaptations to attract insects. Flowering plants also exerted a major influence on the evolution of birds and mammals. Birds, which feed on seeds, fruits, and buds, evolved rapidly in intimate association with the flowering plants. During the Cenozoic era, the emergence of herbivorous mammals coincided with the widespread distribution of nutritious grasses over the plains. In turn, the herbivorous mammals furnished the setting for the evolution of carnivorous mammals. The interdependence between plants and animals continues to exist in nature today.

CONVERGENCE

The major pattern of the evolutionary panorama is divergence—the tendency of a population of organisms to become highly diversified when the population spreads over an area with varied habitats. The population can diverge into radically different lines, each modified for a specific ecological role. Now, essentially similar habitats may be found in widely separate parts of the world. It would thus not be surprising to find that two groups of organisms, unrelated by descent but living under similar environmental conditions in different geographic regions, can exhibit similarities in habits and general appearances. The tendency of one group of organisms to develop superficial resemblances to another group of different ancestry is called *convergence*. Stated another way, convergence is the result of similar selective forces causing common adaptive features of geneticlly independent origin. Contrarily, similarities of features in two groups due to a shared common genetic ancestry constitutes *homology.*

Convergence is not an uncommon phenomenon in nature. Many unrelated, or remotely related, organisms have converged in appearance as a consequence of exploitation of habitats of similar ecological makeup. The relationship of convergence to adaptive radiation should be evident; the former is the inevitable result of the countless series of adaptive radiations that have taken place in scattered parts of the globe. A striking example of convergence is afforded by the living marsupials of Australia.

The massive island continent of Australia has long been isolated from Asia, as least since the commencement of the Cenozoic era, the last 65 million years. It is on this isolated island that the marsupials—primitive pouched mammals—survived, free from the competition of the more efficient placental mammals when these came into prominence. In the absence of a land migration route between Asia and Australia, the latter land mass was inaccessible to practically all the placental mammals. On other continents, the marsupials perished, save for the peculiar American opossums, the didelphids.

Imprisoned in Australia, the marsupials spread into a variety of habitats. Several live in the open plains and grasslands; some are tree-dwellers; others are burrowers; and still others are gliders (fig.16.6). Most kangaroos are terrestrial, but one variety, the monkeylike kangaroo, is arboreal. The slow moving, nocturnal "teddy bear," or koala, lives and feeds on *Eucalyptus,* the dominant tree of Australia. The bandicoot, with rabbitlike ears, has sturdy claws adapted for digging in the ground in search of insects. Marsupial moles live in desert burrows, and flying phalangers have webs of loose skin stretched between the forelimbs and hindlimbs. The flying phalanger cannot actually fly but is adept at gliding. The impressive diversity of marsupials thus represents an admirable example of adaptive radiation.

The marsupials of Australia also vividly illustrate the phenomenon of convergence. They have filled the ecological niches normally occupied by placental mammals in other parts of the world. The marsupial "mouse," "mole," "anteater," "wolf," flying phalanger, and groundhog-like wombat strikingly resemble the true placental types—mouse, mole, anteater, wolf, flying squirrel, and groundhog, respectively—in general appearance and in ways of life.

It is interesting to note that a marsupial "bat" has not evolved in Australia. The opportunity was apparently denied by the invasion of placental bats from Asia, one of the few placental forms that managed, probably as a result of dispersal by flight and winds, to reach Australia. With the coming of

Figure 16.6 Marsupial, or pouched mammals, of Australia, illustrating the twin themes of adaptive radiation and convergence. The marsupials have radiated, or diversified, into a variety of forms, ranging from the tiny insect-eating "mouse" to the fierce, flesh-eating Tasmanian "wolf." Many of the marsupials resemble in appearance and habit the placental mammals of other parts of the world, although thay are not closely related to them.

humans, the secure existence of the marsupials has been threatened. Prehistoric humans introduced dogs, which ran wild (the dingos); later human settlers brought a number of European placental mammals, such as the rabbit, hare, fox, and Norway rat. Among the marsupials faced with extinction are the marsupial "wolf," which survives today only in Tasmania, and the slow moving "anteater," which is rapidly disappearing.

THE PLAINS-DWELLING MAMMALS

When the great majority of land mammals first arose, South America was just as much an island continent as Australia. An impressively rich assemblage of mammals evolved in isolation in South America during the 70 million years that the continent remained separated from the rest of the world. Such bizarre placental mammals emerged as the giant ground sloth, the enormous toothless anteater, and the armored armadillo-like glyptodont. The predatory wolves that evolved in South America were not true carnivores; rather, they resembled the predaceous marsupials of Australia.

The spread of grasses across the flatlands of the Americas in the Tertiary period was associated with the rapid expansion of herbivorous hoofed mammals (ungulates) in both continents. The crowd of grazing mammals that arose is so bewildering that it is difficult to call up a true picture of the immense variety of ungulate types. These archaic hoofed mammals are quite unfamiliar to most people; they have no vernacular names and taxonomists have had to create special suborders for many of them. Selected examples of these curious hoofed mammals were shown earlier in figure 2.3 (chapter 2). These South American forms bear only superficial resemblances to the hoofed mammals in other parts to the world. *Toxodon* superficially resembles a rhinoceros, *Thoatherium* has the appearance of a horse, *Pyrotherium* may be likened to an elephant, and *Macrauchenia* looks like a camel. The similarities in exter-

nal appearances signify convergence and not relationship by descent.

At the time the Isthmus of Panama rose from the depths, the land bridge across the Bering Straits lay completely dry. Large mammoths streamed in increasing numbers across the Bering Strait from Asia into North America during the Pleistocene. The mammoths, particularly the 10-foot high woolly mammoth, dominated the American plains. North America was also the recipient of some of the South American mammals. Among the immigrants from south of the border were *Boreostracon*, a giant glyptodont, encased in armor and an offshoot of the armadillo line, and *Megatherium*, a great ground sloth that weighed more than a modern elephant (fig.16.7).

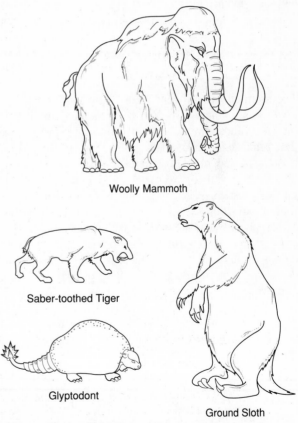

Woolly Mammoth

Saber-toothed Tiger

Glyptodont

Ground Sloth

Figure 16.7 Prominent land mammals that occupied the flatlands of the Americas in the Tertiary period.

The South American hoofed mammals, evolving independently of mammalian types in North America, thrived for many millions of years. Then the water gap between the Americas was closed. The Isthmus of Panama rose above the sea, establishing a land corridor linking the two continents. Through this land bridge infiltrated a horde of northern invaders. Predatory carnivores, such as saber-toothed tigers and wolves, entered South America from the north, as did progressive herbivorous ungulates, among which were tapirs, peccaries, llamas, and deer. All of the native South American hoofed mammals, without exception, were driven into extinction. South America also became deprived of its indigenous variants of ground sloths and glyptodonts. It then later lost several of the immigrants, notably the mastodons and saber-toothed tigers.

EXTINCTION AND EVOLUTIONARY STABILITY

The multitude of different kinds of present-day organisms is impressive. Yet, the inhabitants of the world today are only a small percentage of the tremendous array of organisms of earlier periods. Indeed, roughly 99 percent of all species that have ever lived are extinct. Evidently, the fate of most lineages of organisms throughout time is extinction.

Apparently, only those populations that can continue to adapt to changing environmental conditions avoid extinction. Yet, there is more to the demise of a species than mere failure of adaptation. We now acknowledge the importance of *mass extinctions,* which are the consequences of catastrophic events and occur randomly, without regard to the adaptive values of traits of organisms. Mass extinctions may be the result of cataclysmic earthly events, such as volcanic eruptions, climatic changes, and changes in sea levels. Extraterrestrial events, such as a comet or meteor striking the earth's surface, have also been implicated. The mass extinction of the dinosaurs before the close of the Mesozoic era has been attributed to the impact of a meteorite in the vicinity of the Yucatan peninsula.

The history of biologic evolution on earth has been punctuated by periodic mass extinctions. On each occasion, the massive and rapid extinction of one group of organisms has been followed by an adaptive radiation (diversification) among the survivors. As mentioned previously (chapter 13), the catastrophic departure of the dinosaurs provided evolutionary opportunities for the small and inconspicuous mammals. It may be that the small body size of the early mammals enabled them to survive the catastrophic event when dinosaurs did not. Parenthetically, it may be noted that human-induced extinctions fall in the category of mass extinctions.

Some types of organisms have not changed appreciably in untold millions of years. The opossum has survived almost unchanged since late Cretaceous, some 75 or more million years ago. The horseshoe crab, *Limulus,* is not very different from fossils uncovered some 500 million years ago. The maidenhair, or ginkgo, tree of the Chinese temple gardens differs little from its ancestors 200 million years back. The treasured ginkgo has probably existed on earth longer than any other present tree. Darwin called the ginkgo "a living fossil."

Long-standing stability of organization seems antithetical to the concept of evolution. Nevertheless, the notion that an established species may be rather stable is not completely at variance with conventional evolutionary thought. Long periods of relatively little change may be the outcome of stabilizing selection (see chapter 6), in which the typical members of a population are selectively favored relative to the extremes. In other words, the "status quo" is maintained by selection after an ecological niche in a habitat has been successfully filled. Admittedly, to some investigators, it seems implausible that natural selection could be so conservative as to continue to favor the same phenotype for many millions of years.

EMERGENCE OF THE HUMAN SPECIES

The human species is the outcome of the same natural forces that have shaped all animal life. As relatively recent arrivals on the earth, humans represent one of the latest adaptive advances in the animal kingdom. We are clearly part of the fabric of nature and have ancestral ties with other animals. There is almost universal unanimity that our closest relatives are the apes. The line leading ultimately to humans diverged from the ape branch during Tertiary times. There are several candidates among Tertiary fossil apes that qualify as ancestors of the human species. One often hears stated, hesitatingly and perhaps in the form of an apology, that we are not really closely related to the apes but merely share a very distant ancestry with them. By indirection, the idea is conveyed that our remote generalized ancestor would not be at all apelike. This is sheer deception. There is simply no escaping the fact that our very early predecessors had many features in common with the apes.

PRIMATE RADIATION

Primates, the order to which humans belong, underwent adaptive radiation in Cenozoic times,

approximately 70 million years ago. The primates are primarily tree dwellers; only humans are fully adapted for life on the ground. Many of the noteworthy characteristics of the primates evolved as specializations for an arboreal mode of life. Depth perception is important to a tree-living animal; the majority of the primates are unique in possessing binocular, or stereoscopic, vision, wherein the visual fields of the two eyes overlap. The hands evolved as prehensile organs for grasping objects, an adaptation later useful to humans for manipulating tools. The use of the hands to bring objects to the nose (for smelling) and to the mouth (for tasting) was associated with a reduction of the snout, or muzzle, and a reduction of the olfactory area of the brain. Primates generally have a poor sense of smell. Closely associated with enhanced visual acuity and increased dexterity of the hands was the marked expansion of the brain, particularly the visual and memory areas. Progressive enlargement of the brain culminated, in humans, in the development of higher mental faculties.

The primate stock arose and differentiated from small, chisel-toothed, insectivorous ancestors (fig. 17.1). The Asiatic tree shrew, *Tupaia,* an

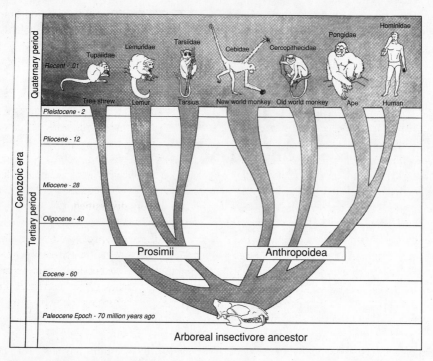

Figure 17.1 **Adaptive radiation of primates** from a basic stock of small, insect-eating placental mammals, the Insectivora (whose living kin include the shrews, moles, and hedgehogs). Based on paleontological and anatomical evidence, the pongid (ape) and hominid (human) lineages diverged during the Miocene epoch, about 20 million years ago. Estimates based on biochemical data are presently at variance with the fossil data. The biochemical evidence suggests that the lineage leading to the human split off from the ape line only about 5 million years ago.

agile tree climber that feeds on insects and fruits in tropical forests, survives as a model for the ancestor of the primates. Most authorities place the tree shrew in the mammalian order Insectivora, while a few hold the view that the tree shrew is a primitive, generalized member of the primates. The important consideration is that the modern tree shrew bears likeness to the common ancestor of primates.

The order Primates is generally divided into two groups or suborders, the *Prosimii* and the *Anthropoidea.* The prosimians are arboreal and largely nocturnal predators; they include such tropical forms as the lemurs, the lorises, and the tarsiers. As adaptations to an arboreal-nocturnal niche, these prosimians evolved large, forwardly placed eyes and

strong, grasping hands and feet. The lemurs are largely confined to the island of Madagascar, the lorises are found principally in eastern Asia, and the tarsiers are limited to southeastern Asia. The tarsiers represent the most advanced group of prosimians and may be said to foreshadow the higher primate trends of the Anthropoidea.

The more advanced primates, the anthropoids, are composed of the New World monkeys, the Old World monkeys, the apes, and humans. All are able to sit in an upright position, and thus the hands are free to investigate and manipulate the environment. The monkeys are normally *quadrupedal,* running on all fours along branches of trees. Among the apes, gibbons habitually use their arms for hand-over-hand swinging in a motion known as *brachiation.* The great apes (orangutan, chimpanzee, and gorilla) can maintain prolonged semi-erect postures. When on the ground, the chimpanzee and gorilla are "knuckle-walkers," the weight being placed on the knuckles of the hands rather than on the extended fingers. The chimpanzee and the gorilla regularly travel quadrupedally on the ground. Humans alone are specialized for erect bipedal locomotion.

Recent biochemical and genetic studies reveal that the two species of chimpanzees (the common chimpanzee, *Pan troglodytes,* and the bonobo—once called the pygmy—chimpanzee, *Pan paniscus*) and the three subspecies of gorillas (western lowland *Gorilla gorilla gorilla,* mountain *Gorilla gorilla beringei,* and eastern lowland *Gorilla gorilla grauer*) bear a special evolutionary relationship to humans. A wealth of molecular data (see chapter 15) reveals that the African great apes (chimpanzee and gorilla) are more closely related to humans than they are to Asian apes (orangutans). Some taxonomists firmly maintain that the molecular findings should be reflected in our classification scheme *by placing the African apes in the family of humans* (Hominidae). By one proposed revision, the Hominidae would be divided into subfamilies Paninae (chimpanzee and gorilla) and Homininae (humans). The traditional division of Pongidae (chimpanzees, gorillas, and orangutans) and Hominidae (humans) is still the most widely used, even though it may fall short of representing the true state of affairs.

ADAPTIVE RADIATION OF HUMANS

The morphological differences between humans and apes relate mainly to locomotory habits and brain growth. Humans have a fully upright posture and gait and an enlarged brain. The cranial capacity of a modern ape rarely exceeds 600 cubic centimeters, while the average human cranial capacity is 1,350 cubic centimeters. The mastery of varied environments by humans has been largely the result of superior intelligence gradually acquired throughout evolution.

It may be difficult to envision humans other than as they exist today. Nevertheless, it should not be imagined that fossil specimens of the early Pleistocene epoch possessed the attributes of contemporary forms. Present-day humans are certainly different from their predecessors, in much the same manner that any modern form of life is different from its forerunner. It is important to recognize that there have been different kinds of human ancestors. The evolutionary process of adaptive radiation produced a family of human ancestors in much the same way that other groups of organisms became highly diversified.

Our present-day species—*Homo sapiens*—is the sole surviving species of human. Until recently, we conceived of the human lineage as a tidy, straightforward succession of one species to the next, leading eventually to modern *Homo sapiens.* We scarcely permitted ourselves to evoke the proposition that two or more species could have *overlapped in place or time.* The saga of human evolution increasingly suggests that there were *coexisting species of human ancestors* throughout most of our evolutionary past.

FORERUNNERS OF THE GREAT APES

The modern apes are but a remnant of a once diverse and widespread group. The great apes reached their evolutionary peak in the Old World forests of the early Miocene, some 20 million years ago. The most famous of the Miocene apes has been classified as *Dryopithecus,* the oak-ape, so called because of the presence of oak leaves in the fossil beds. Known in more informal terms as the dryopithecines, these primitive oak-apes were a relatively successful group of 30 or more species that flourished for at least 10 million years. At that time, the Arabian peninsula joined Africa and Eurasia together, and the dryopithecines ranged widely throughout the two great land masses. Then, after their remarkable spread and divergence, nearly all of these early apes became extinct.

The African variety of ape, *Dryopithecus africanus*—first named *Proconsul,* after a famous chimpanzee at the London Zoo—may not be far removed from the common stock from which apes and humans arose. Although primarily a tree dweller, *Dryopithecus africanus* apparently wandered on the ground and profited from the vegetable foods in the open grasslands (fig. 17.2). The transition from tree dwelling to ground living might well have first appeared at this time. The Miocene epoch was characterized by expanding populations of varied species of monkeys and apes in the African forests. Population pressure probably contributed to the exploratory departure of oak-apes from the forest fringes to the savanna grasslands. These pioneer ground-dwelling apes

Figure 17.2 Reconstruction of *Dryopithecus aficanus,* an apelike type that prowled East Africa 20 to 25 million years ago. This pongid apparently led an agile life both on and off the ground.

(Painting by Maurice Wilson; by permission of the Trustees of the British Museum—Natural History.)

apparently were able to establish themselves in the open bushy plains by subsisting mainly on plant foods and largely avoiding confrontations with the savanna carnivores. But as the forests continued to recede and yield to the grasslands, the dryopithecines became extinct. Some species of this ape may have evolved into the still surviving orangutan, gorilla, and chimpanzee.

The modern apes are intimately tied to the steadily shrinking forest habitats and are likely to face imminent extinction in the wild. The population numbers of the great apes have dwindled because the apes have failed to adapt to niches beyond the humid tropical forests. Contrarily, the monkeys have thrived as they have successfully undergone adaptive radiation into diverse habitats.

FORERUNNERS OF THE HOMINIDAE

Included in the great Miocene diversity of apes in Africa and Eurasia were two enigmatic apes with hominid-like dentition. These apes were named *Ramapithecus* and *Sivapithecus*, after the Hindu deities Rama and Siva. Ever since the first fragile upper jaw of a ramapithecine was found by G. Edward Lewis in the early 1930s in the Silawik Hills of northern India, anthropologists have argued over whether these fossils were the first ancestral hominid, an ancestral pongid (in particular, a proto-orangutan), or merely an evolutionary dead end.

The striking ramapithecine fossil jaws exhibit reduced canine teeth, relatively small incisors, and low-crowned molars capped with a thick layer of enamel. The thickly enameled molar teeth suggest that *Ramapithecus* ate chiefly hard-covered fruits, nuts, and seeds rather than the softer forest fruits usually consumed by apes. Additionally, the flat wear of the molars hints of a grinding type of mastication. Such grinding action would be unusual for apes, who tear and crush, rather than masticate, their succulent food.

When the jaw of *Ramapithecus* was originally reconstructed, the dental arcade was presumed to be parabolic in shape, as in hominids (fig.17.3).

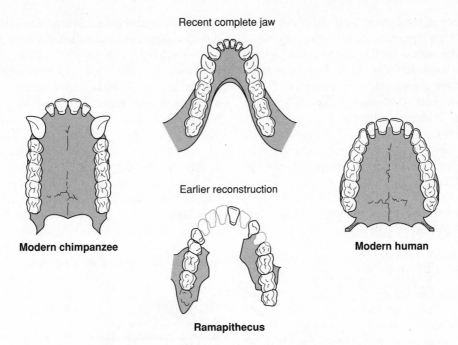

Figure 17.3 **Comparison of dental arcades** of chimpanzee *(left),* human *(right),* and *Ramapithecus (middle).* In a once-accepted reconstruction of the jaw of *Ramapithecus,* the shape was concieved as parabolic, like that of humans. Recent evidence indicates that the dental arcade of *Ramapithecus* is V-shaped. Note that the long, projecting canines of the chimpanzee contrast with the relatively shortened human canines.

The smoothly rounded contours of the arcade suggested that *Ramapithecus* had, in incipient form, the dental and facial organization characteristic of later hominids. In the 1960s, Elwyn Simons of Yale University championed the view that *Ramapithecus* may be close to the base of the stem of the hominid family. That is to say, this fossil form marks the point where the hominid lineage separated from the pongid assemblage. In fact, it became an acceptable view that *Ramapithecus* was the earliest member of the human family (Hominidae). If this view were valid, it would place the first discernible human origins at about 16 million years ago.

In the last several years, there has been a marked inclination to dethrone *Ramapithecus* as a hominid. With the discovery of complete jaws of

Ramapithecus, it was found that the dental arcade was unmistakably V-shaped, rather than parabolic (fig. 17.3). Adrienne Zihlman of the University of California, Santa Cruz, assesses *Ramapithecus* as merely one of several kinds of apes that ventured into open country more than 14 million years ago. The subsequent discovery of fossilized limb bones from Pakistan suggests that *Ramapithecus* was more quadrupedal than previously thought and more arboreal than terrestrial. The Harvard anthropologist David Pilbeam favors the view that *Ramapithecus* and *Sivapithecus* represent groups that were ancestral to the orangutan, which split off from the lineage leading to hominids about 16 million years ago.

If the apelike *Ramapithecus* is on the line to the common ancestry of both apes and humans, it is most likely distantly removed from the actual pongid/hominid divergence. As previously mentioned (chapter 15), data from molecular studies suggest that the bifurcation between pongids and hominids occurred 4 to 6 million years ago. Investigators who persist in advocating the hominid status of *Ramapithecus* dismiss the molecular evidence by claiming that the molecular clock is not finely tuned and keeps inaccurate time. However, most evolutionists have removed *Ramapithecus* from its once distinguished position as our oldest ancestor.

AUSTRALOPITHECINES: THE FIRST HOMINIDS

There is, at present, an inexplicable 3-million-year gap between the disappearance of *Ramapithecus* about 7 million years ago and the arrival of the first undeniable hominid—*Australopithecus*— about 4 million years ago. The australopithecine stage is a relatively long phase of adaptive evolution, a phase in which several species of early hominids apparently coexisted. The australopithecine era commenced 4 million years ago and ended about 1 million years ago.

In 1924, the Australian anatomist, Raymond A. Dart, announced the discovery of an unusual small skull from a Pleistocene limestone quarry near the village of Taung in the Transvaal region of South Africa. The fossilized skull was that of a child of about six years. The little Taung skull bears some resemblance to the skull of a young chimpanzee, but many of its components, notably the teeth, show pronounced affinities to humans. Certain striking features of the Taung skull suggest that the child had walked upright. The remarkable skull was designated by the formidable name of *Australopithecus africanus* (*austral,* for "south," *pithekos,* for "ape," and *africanus,* for "from Africa"). Dart was confident that *Australopithecus* was related to the ancestral stock of humans rather than to the great apes.

Dart's declaration, which several scientists initially derided, was fortified by findings in the 1930s by the late Robert Broom, a Scottish pale-

ontologist. Adult skulls of *Australopithecus* were dug out from caves in Sterkfontein, Kromdraai, and Swartkrans in South Africa. The adult skulls confirmed the hominid anatomical pattern seen in the juvenile Taung cranium. The several new fossil forms were originally given different names. However, in recent years, it has become customary to refer to the South African fossils collectively as the australopithecines. They were short, four to five feet in height, with a small ape-sized brain (cranial capacity range of 450 to 600 cubic centimeters). Nonetheless, the australopithecines stood upright, walked bipedally, and dwelt in open country (fig. 17.4). These circumstances nullify the popular view that humans were intelligent animals when they first came down out of the trees. It

Figure 17.4 **Reconstruction** of *Australopithecus,* a hominid of about 4 million years ago who stood upright, walked bipedally, and dwelt in open country.
(Painting by Maurice Wilson; by permission of the Trustees of the British Museum—Natural History.)

seems clear that erect bipedal locomotion on the ground preceded the development of a large complex brain. Moreover, the upright posture conflicts with popular iconography that typically illustrates a hunched-over apelike ancestor that becomes progressively upright over a stretch of time.

The australopithecines are decidedly early representatives of the hominid lineage. However, it is not clear which of the australopithecines occupies a prominent place in the *direct* ancestry of humans. The initial fossil findings fostered the notion that there were at least two distinct species of australopithecines—the light-jawed, slender ("gracile") *Australopithecus africanus* and the heavy-jawed, robust *Australopithecus robustus* with extremely large grinding molars (fig. 17.5). The former species was not over four feet tall and weighed no more than 60 pounds, whereas the latter species was a foot taller and at least 30 pounds heavier. Ecologists have observed that when two contemporaneous species occupy the same habitat, the potential competitors become differentially specialized to exploit different components of the local environment. Thus, direct competition for food resources is minimized and the two species are able to coexist. It is thought that *Australopithecus africanus* increasingly supplemented its diet with animal food. The dietary difference is supported by the finding that *A. africanus* had smaller molars than *A. robustus.*

If, as it appears likely, the two *Australopithecus* species did coexist at the same time in the same region, then only one of the two could have been the progenitor of a more modern species of hominid. Paleontologists had earlier suggested that the vegetarian *A. robustus* perished without leaving any descendants, and that *A. africanus* was the forebear of a more advanced hominid. This was based on the assumption that a dietary change to carnivorism represented one of the more

important steps in transforming a bipedal ape into a tool-making and tool-using human. Current thinking does *not* place *A. africanus* in the direct line of ancestry of humans (fig. 17.5).

The South African findings have been supplemented and extended by fossil discoveries by the late Louis Leakey and his wife Mary at the 25-mile-long Olduvai Gorge in Tanzania. Olduvai Gorge is situated in a volcanically active region in

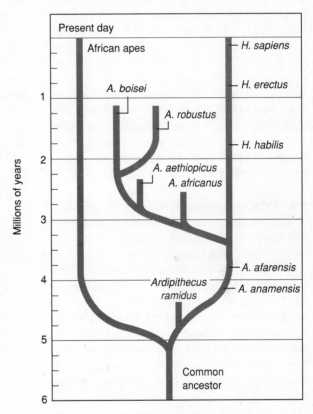

Figure 17.5 Elemental phylogeny of humans. In this rudimentary scheme, *Australopithecus afarensis* followed by *Homo habilis* and subsequently *Homo erectus* are depicted as evolving successively in geologic time, forming a straightforward lineage leading to *Homo sapiens*. In the distant past, an array of australopithecine species coexisted with the *Homo* assemblage. The recently discovered, chimplike *Ardipithecus ramidus* has been claimed to be the root or base of the hominid family tree. Compare this rendition of the evolutionary tree with the portrayal in figure 17.9.

(From P.H. Raven and G.B. Johnson, *Biology*, 4th ed. Dubuque, IA: WCB/McGraw-Hill, 1996, fig. 23.8.)

eastern Africa and has invaluable sediments of lava from prehistoric volcanic activity. Since lava contains argon 40, the daughter isotope of potassium 40, the rate of decomposition from potassium to argon permits an accurate dating of the Olduvai layers. Like a miniature Grand Canyon, the sides of Olduvai Gorge display different strata laid bare by the cutting of an ancient river.

In an exposed stratum of the gorge, the Leakeys in 1959 uncovered bony fragments of a robust australopithecine, characterized by extremely massive jaws. This heavy-jawed fossil form with enormous teeth was called *Zinjanthropus,* or the Nutcracker Man. Fossil remains of *Zinjanthropus* were found in strata judged, by the potassium-argon dating method (instead of the conventional uranium-lead technique), to be about 1.8 million years old. Most scientists today agree that *Zinjanthropus* is essentially an eastern African variety of *Australopithecus robustus.* Others contend that *Zinjanthropus* has more exaggerated, or coarser, features than *A. robustus* and warrants recognition as a separate species, *Australopithecus boisei.* It would thus represent yet another species in the australopithecine complex.

One might suppose than an older, as yet undiscovered australopithecine (from Pliocene sediments) was at the base of the hominid lineage. The candidate for such an ancestor was thought to be the fossil hominid recovered in 1974 by a team led by the anthropologist Donald C. Johanson, now at the Institute of Human Origins in Berkeley. The remains of this hominid were found at a formation in the Hadar region of Ethiopia, a formation which has been dated at 3.2 to 3.5 million years ago. The fossil beds at Hadar have yielded a remarkably rich paleontological collection. The constellation of fossils includes a social group of 13 individuals, a group that has come to be known as "the First Family." The family members apparently died together at one spot. The most complete adult skeleton uncovered is the now famous female hominid called "Lucy." Her name was borrowed from the Beatles' song "Lucy in the Sky with Diamonds." She was only 3.5 to 4.0 feet tall, weighed less than 66 pounds (30 Kg), and died when she was in her early twenties (fig. 17.6). Other Hadar individuals are relatively large; the variation perhaps reflects differences due to sex. Parenthetically, it may be noted that an analysis of Lucy's pelvis casts doubt on whether she was a female!

In 1978, Mary Leakey discovered a series of footprints made by human ancestors 3.6 million years ago at Laetoli in northern Tanzania. Apparently, the nearby volcano Sadiman experienced a mild eruption followed by a rainstorm just before at least two hominids walked across this wet cementlike surface. When the volcano later covered the footprints with additional volcanic ash, the footprints were preserved. After the sediments eroded, the footprints were re-exposed. The Laetoli footprints are remarkably similar to modern human feet. Evidently, upright bipedalism was established as early as 3.6 million years ago.

Johanson contends that the hominid remains from Hadar and Mary Leakey's Laetoli fossils are so similar morphologically as to belong to a single

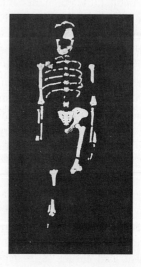

Figure 17.6 **"Lucy,"** a small adult female (less than 4 feet tall) who lived close to 3 million years ago. This famous skeleton was discovered by the paleoanthropologist Donald Johanson, and placed in a new species, *Australopithecus afarensis.*

(Photograph courtesy of Peter Schmid, PhD. Appeared in E. P. Volpe, *Biology and Human Concerns,* 4th ed., fig. 40.6.)

human lineage. In 1978, Johanson and Tim D. White (an anthropologist from the University of California at Berkeley) assigned the Hadar and Laetoli specimens to the same species and delineated them as a new species, *Australopithecus afarenis*. The new species derives its name from the Afar locality in Hadar, Ethiopia, which yielded the most numerous specimens. *Australopithecus afarensis* was then pronounced as the root stock from which all later hominids sprang. At present, it is safe to say that the *afarensis* remains, dated between 3 and 4 million years ago, constitute one of the earliest definitive members of the family Hominidae (fig. 17.5). There are critics who contend that the bones and teeth of the Hadar individuals provide evidence for two distinct species. If true, this would push the search for a common ancestor back even further in time.

Indeed, as might have been anticipated, older and more primitive hominids dating between 4.2 and 4.5 million years ago have been unearthed. Among them, as seen in figure 17.5, is *Ardipithecus ramidus* (*Ardipithecus* means "ground ape" and *ramidus* means "root"), whose remains were found at the Ethiopian site of Aramis. As the species name implies, this fossil has been proclaimed to be at the root or base of the human lineage. *Ardipithecus ramidus* may be the proverbial "missing link" between chimpanzees and humans.

In essence, an enormous burst of diversity characterized the early stages of hominid evolution. There were evidently multiple, coexisting species of hominids. The challenge is to trace the line of descent of our own genus, *Homo*, from the broad australopithecine family tree.

HUMANS EMERGE: *HOMO HABILIS*

Members of our genus, *Homo*, evolved in Africa from australopithecine ancestors approximately 2.5 million years ago. We are continually gaining new insight into how the genus *Homo* evolved and how its members came to live with, and eventually replace, the australopithecines. The increasingly diverse fossil record permits us to portray human evolution as a flourishing bush containing various species. Although only a solitary human species *(Homo sapiens)* exists today, several species coexisted for much of our history.

Early in the 1960s, Louis and Mary Leakey found remains of a light-jawed hominid that they claimed was more advanced and more humanlike than any known australopithecine. Estimates of the brain capacities of the skulls averaged 637 cubic centimeters. This light-jawed type of Olduvai Gorge hominid was said to be the first civilized or humanized being, deserving of the rank of *Homo,* namely *Homo habilis* (fig. 17.5). The specific name *habilis* means "handy," from the inferred ability of this hominid to make stone tools. The recognition of *Homo habilis* indicates that this primitive human being was evolving alongside the less hominized australopithecines and lived side by side with them. The coexistence of *Homo habilis* and the australopithecines is generally accepted.

Spirited discourse surrounds son Richard Leakey's 1972 discovery of an unusual skull from the desiccated fossil beds of Lake Turkana in Kenya. This nearly complete skull was initially placed at 2.8 million years old but is currently acknowledged as being only 1.8 million years old. The specimen was cautiously designated only by its museum identification number—"1470." The cranial capacity of the "1470" skull measures 780 cubic centimeters, which is significantly larger than any australopithecine specimen. Skull "1470" clearly belongs to the genus *Homo.* Some anthropologists believe that "1470" is sufficiently different from *Homo habilis* to have earned its own species name, *Homo rudolfensis.* In the most current configuration, *Homo rudolfensis* is viewed as a distinct species that evolved earlier than *Homo habilis.*

GATHERING-HUNTING WAY OF LIFE

The lower Pleistocene hominids *(Homo habilis)* fabricated crude chopping tools by striking a few flakes from a cobble or large pebble. The stone tools enabled *habilis* to cut scavenged meat into chewable size pieces, as well as smash large bones to get at the fat-rich bone marrow. The open African savanna offered the early plains-dwelling hominids a wider selection of foods than were available in the tropical forests. Tools were used to gather a variety of plant materials (nuts, seeds, fruits, tubers, and roots) and small animals (lizards and rodents). It is unlikely, as popularly imagined, that tools were first contrived as efficient spears to hunt large game or that our early ancestors were voracious meat-eating, predatory primates. The human species became accomplished hunters of the large animals only later in evolutionary history.

The early hominids on the savanna were most likely primarily gatherers of plants and small animal life, and both males and females were opportunistic in devising tools for digging, processing plant foods, and scavenging meat. The gathering of food was particularly critical to females with dependent young. There may have been some stalking and killing of game, but the uncertainty of success made hunting secondary or supplementary to the collection of plant food with its assurance of success. The exploitation of food sources by both males and females on the savanna fostered the collaborative interactions of individuals. Cooperative behavior promoted the sharing of food for the first time. The sharing of food regularly is a social achievement unique to humans; only rarely do apes share food. The exchange of food is considered to be the earliest expression of human social reciprocity.

Homo habilis apparently had little propensity to migrate beyond Africa. This early human thrived for nearly 500,000 years before it became extinct. The hominid that followed had an even greater brain size and a proclivity to migrate.

HOMO ERECTUS: THE EXPLORER

The earliest and best-known representative of *Homo erectus* is the famous Java fossil, first described as *Pithecanthropus erectus* ("upright ape-man"). This primitive human being was dis-

covered at Trinil, Java, in 1894 by Eugene Dubois, a young Dutch army surgeon. Dubois had been profoundly influenced by the writings of Charles Darwin and was taken with the idea that he could find the origins of humans. He surprised the world with the discovery of this early human. Curiously, Dubois in his later years inexplicably doubted his own findings and contended that *Pithecanthropus erectus* was merely a giant manlike ape. In the 1930s, additional fossil finds of *Pithecanthropus* were unveiled in central Java by the Dutch geologist G.H.R. von Koenigswald. The fossil specimens date from between 1 million years and 700,000 years ago. The newer findings confirmed the human status of the pithecanthropines.

The pithecanthropines lived during middle Pleistocene times, between 1.3 million and 300,000 years ago *or less*! They arose during a period of shifting climates that turned much of forested Africa into cooler and dryer open grasslands. These low-browed hominids were toolmakers and hunters who had learned to control fire. They probably had some powers of speech. Their ability to exploit the environment is reflected in the expanded size of the brain. The cranial capacity of the pithecanthropines was in the range of 775 to 1,000 cubic centimeters. Their advanced tool kit (termed *Acheulean*) included finely worked, teardrop-shaped hand axes, various sharp cleavers, finger-sized scrapers, and cutting flake tools. The pithecanthropines were travelers and explorers that migrated successfully through the continents, from the tropical regions of Africa to Asia and Europe.

Current evidence suggests that at least two species deserve recognition. Many writers now reserve the name of *Homo erectus* for the Eurasian pithecanthropine specimens and accord the name of *Homo ergaster* for African finds. In this scheme, *Homo erectus* evolved from the African *Homo ergaster* and migrated out of Africa about 1 million years ago, if not earlier. In 1996, an interdisciplinary team of scientists re-examined two major fossil sites along the Solo River in Java and found surprisingly that *Homo erectus* apparently persisted in southeast Asia until about 27,000 to 53,000 years ago. If proven correct, this would mean that *Homo erectus* survived some 250,000 years after it had been surmised to have become extinct. This remnant population of *Homo erectus* would have existed at the same time that two other human species, *Homo neanderthalensis* and *Homo sapiens* roamed the earth (fig. 17.7). Confirmation of the dates for the Java specimens would provide support for the new paradigm that more than one hominid species could have existed at any one time-level.

In the 1920s, Canadian anatomist Davidson Black's elaborate excavation of caves in the limestone hills near Peking, China, led to the discovery of another primitive human, *Sinanthropus pekinensis,* or "Peking man" (fig. 17.8). The cranial capacity in the sinanthropines varied from 900 to 1,200 cubic centimeters. They fashioned tools and weapons of stone and bones, and they kindled fire. A strong suspicion was once held that *Sinanthropus* was cannibalistic and savored human brains, for many of the fossil braincases show signs of having been cracked open from below. This view is now considered untenable; the damaged braincases are regarded as the work of

scavengers or other natural events. Most of the original fossil specimens of Peking man were lost during World War II. Fortunately, photographs and measurements were made of the original fossils. Plaster casts made of the original fossils still exist.

The Java and Peking hominids were originally each christened with a distinctive Latin name, *Pithecanthropus erectus* and *Sinanthropus pekinensis,* respectively. There is , however, no justification for recognizing more than the single genus of humans, *Homo.* Accordingly, modern taxonomists have properly assigned both the Java and Peking fossils to the genus *Homo.* Moreover, the morphological differences between these two fossil humans are readily within the range of variation that we observe in living populations today. These forms thus represent two closely related geographic races (subspecies) of the same species. Both Java and Peking hominids are usually placed together and classified as *Homo erectus.*

In 1984, an impressive complete skeleton of an adolescent boy dating from 1.6 million years ago was discovered from Kenya's Turkana basin. Dubbed the "Turkana boy," this young male fossil was 5 feet 3 inches tall and between 9 and 11 years of age. It is estimated that his adult height would

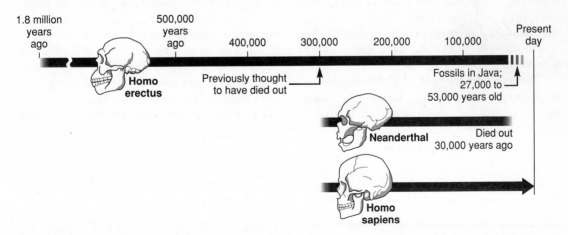

Figure 17.7 **Fossil skulls** found in Java suggest that *Homo erectus,* thought to have perished hundreds of thousands of years ago, coexisted on earth with the Neanderthals and modern *Homo sapiens* as recently as 30,000 years ago.
(Source: New York Times, 12/13/96, front page, *3 Human Species Coexisted on Earth, New Data Suggest* by John Noble Wilford.)

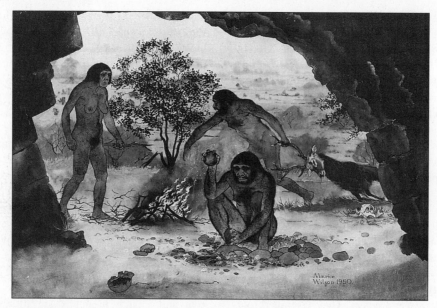

Figure 17.8 Peking humans left remains about 400,000 years ago in limestone caves in northern China, kindled fire, and fashioned tools of stone and bone.

(Painting by Maurice Wilson; by permission of the Trustees of the British Museum—Natural History.)

have been 6 feet, much taller than previous hominids. He possessed a surprisingly modern human body structure, suggestive of modern gait. He was tall, slender, and had long limbs, traits that would appear to be ideally suited for traveling long distances in harsh climactic conditions. This Kenyan specimen has several distinctive features that distinguish it from the Asian *Homo erectus.* Accordingly, the Turkana boy and his kind in Africa have been set apart as a separate species, the aforementioned *Homo ergaster.* This species has been proposed as a credible ancestor for all successive humans.

EVOLUTION OF HUMAN SOCIETY

As *Homo erectus* wandered from place to place, the hunting of game most likely became intensified. Vegetable foods continued to be important in subsistence, but the hunting habit was the way of life for nomadic *Homo erectus.* There is no evi-dence that the female participated in the hunting of large game; the adult female was increasingly encumbered with a fetus or by the care of the young, or by both. The long period of dependency of the young strengthened mother-child bonds but also restricted the mobility and activity of the woman. It appears likely the female remained at the home base as a food gatherer while the male engaged in hunting. The immobility of the female and the prolonged immaturity of the young, cou-pled with the limitations imposed on the male in the number of females he could possibly support, apparently transformed a basically polygamous society into a monogamous structure.

A primary human innovation was relatively permanent pair-bonding, or monogamy. Sustained pair-bonding proved to be advantageous in sever-al respects. It served to reduce sexual competition among the males. The prolonged male-female pairing increased the probability of leaving descendants. The heterosexual pair-bonding rela-tionship became fortified as the estrus cycle of the female became modified into a condition of continuous sexual receptivity. The sustained sex interests of the partners made possible by the obliteration of estrus in the female increased the stability of the family unit and facilitated the development of permanent family-sized shelters for rest, protection, and play. In essence, strong interpersonal bonds between a male, a female, and their children became the basis of the uniquely human family organization. The human male had become incorporated in the mother-young social group of the monkeys and apes.

Human language was fostered as males and females recorded experiences with each other and transmitted information to their mates and children. Speech favored cooperation between local groups and the fusion of small groups into larger communities. Speech also fostered the successful occupation of one geographical area after another. The exchange of ideas over wide areas permitted human cultures of great complexity to develop.

EMERGENCE OF MODERN HUMANS

One might expect that the fossil record of the last 300,000 years would present an unmistakably well-defined picture of the evolution of our modern species, *Homo sapiens.* This is not the case. Like a crafty mystery thriller, the plot of human evolution has been thickened by beguiling fossil characters. The final stages leading to contemporary humans present several unresolved problems. We have yet to disentangle the transitional steps from the *Homo erectus* level of humans to the *Homo sapiens* grade. Moreover, the grade of *sapiens* contains at least two contrastingly different anatomical types—the bulkily built and heavily muscled Neanderthals and the slim-bodied Cro-Magnons. Another fractious issue is the status of *Homo antecessor,* a species newly described in 1997 from lower Pleistocene remains in Spain. *Homo antecessor* has been promulgated as the last common ancestor of the Neanderthals and Cro-Magnons. This rewriting of the script would effectively displace the highly regarded *Homo erectus* from the main line of descent leading to modern humans. Figure 17.9 represents the attempt to place the cast of fossil hominids in their appropriate roles. The number of question marks serves to suggest that re-casting will take place as more fossil characters are added.

STATUS OF THE NEANDERTHALS

The classic Neanderthal was first unearthed in 1856 in a limestone cave in the Neander ravine near Düsseldorf, Germany (fig. 17.10). This group of fossils derives its name from the picturesque Neander Valley or Neanderthal ("thal" means valley, but the silent "h" has been dropped in the modern German to "tal"). The Neanderthals or Neandertals (both spellings continue to be used) are one of the best known of fossil hominids, having been found at numerous widely separate sites in Europe, particularly in France. The Neanderthals were cave dwellers, short (about 5 feet) but powerfully built, with prominent brow ridges. They had large brains with an average capacity of 1,450 cubic centimeters, as opposed to 1,350 cubic centimeters in modern humans. The Neanderthals occupied Europe, the Middle East, and western Asia from at least 230,000 years ago until about 30,000 years ago. Their most immediate ancestor is presumed to be *Homo heidelbergensis* (fig. 17.9).

The Neanderthals were once popularly portrayed as brutish and dull-witted human creatures. On the contrary, the Neanderthals made complex stone tools, were accomplished hunters of large game, compassionately cared for their sick and infirmed, and withstood the rigors of the bitter cold climate of the last glaciation. There is evidence that the Neanderthals buried their dead with various ritual objects. The tools of the Neanderthals are known as *Mousterian* and are characterized by flint scrapers and points. The Neanderthals roamed over Europe during the Upper Pleistocene until about 30,000 years ago and then dramatically disappeared. They were

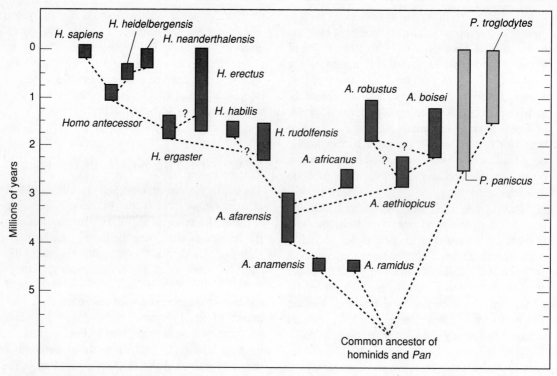

Figure 17.9 **Adaptive radiation of humans.** This portrayal presents the view that our human ancestors underwent extensive speciation and that there were coexisting species of human ancestors throughout most of our evolutionary past. One of the more recent provocative theses is that the newly discovered *Homo antecessor* is the last common ancestor of the Neanderthals and *Homo sapiens*. *Homo antecessor* displaces previously favored contenders—*Homo erectus* and *Homo heidelbergensis*—from the main line of descent leading to modern humans.

(Modified from P.H. Raven and G.B. Johnson, *Biology*, 4th ed. Dubuque, IA: WCB/McGraw-Hill, 1996, fig. 23.9.)

replaced by humans of a modern type, much like ourselves, which have been grouped under the common name of Cro-Magnon. The Cro-Magnons are decidedly representatives of our own species, *Homo sapiens*.

The origins and affinities of the Neanderthals are problematic. Were the Neanderthals our immediate ancestor, or were they sterile offshoots of the human family tree? Some investigators have proposed that the Neanderthals be treated as different from modern *sapiens* only at the subspecies level. Thus, the Neanderthals are to be classified as *Homo sapiens neanderthalensis*. Under this view, the taxonomic title of the Cro-Magnons would be *Homo*

sapiens sapiens. Increasingly, however, researchers are according the Neanderthals their own species status, *Homo neanderthalensis*. By this view, both our species, *Homo sapiens,* and the Neanderthals *(Homo neanderthalensis)* are descended from a common ancestor—perhaps *Homo heidelbergensis* or the recently described *Homo antecessor* (fig. 17.9). In essence, most anthropologists regard the Neanderthals as an evolutionary dead end and not as our immediate ancestor.

There is evidence that modern humans displaced the Neanderthals in the Middle East by 40,000 years ago. Modern humans spread into Europe about 40,000 years ago. From 40,000 to

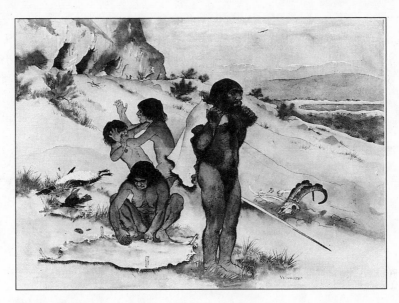

Figure 17.10 Neanderthal, a rugged cave-dweller who roamed Europe and the Middle East about 75,000 years ago.
(Painting by Maurice Wilson; by permission of the Trustees of the British Museum—Natural History.)

30,000 years ago, a span of 10,000 years, the Neanderthals and Cro-Magnons appear to have lived side by side in Europe. Some authors have suggested that the Cro-Magnons conquered and destroyed the Neanderthals. Others have endorsed the view that the anatomically modern human migrants brought new diseases to Europe for which the Neanderthals had no resistance. There is also the opinion that modern humans and the Neanderthals interbred and assimilated their respective gene pools. This opinion apparently has been dispelled by mitochondrial DNA studies, which reveal that the Neanderthals became extinct without contributing mitochondrial genes to modern humans.

MODERN HUMANS: THE CRO-MAGNONS

The Cro-Magnons, a representative of our own species, *Homo sapiens,* can be traced back about 100,000 years ago in Africa and the Middle East, about 50,000 years ago in Australia, and about 40,000 years ago in Europe (fig. 17.11). The Cro-Magnons were tall and slender: males reached 6 ft. (1.8 m) and weighed 155 lbs. (70 kg) while females reached 5 ft. 6 in. (1.7 m) and weighed 120 lbs. (55 kg). The Cro-Magnons were culturally superior to the Neanderthals. Their refined *Aurignacian* tool kit included improved hunting weapons (spears, nets, harpoons, and hooks), elaborate clothing from animal skins and furs, and sophisticated art and culture (fig. 17.12). Magnificently engraved bones, paintings, and sculptures have been recovered from caves in Spain and France. Modern-day artists view with awe the brilliant Cro-Magnon paintings and carved statues, often located deep within caves.

PLEISTOCENE OVERKILL

The Cro-Magnons were accomplished hunters, so much so that they have been held responsible for the high rate of destruction and extinction of the mammalian fauna—such as the giant sloth,

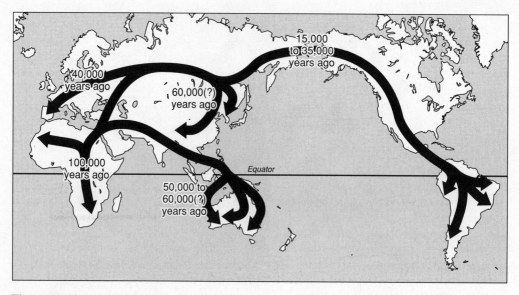

Figure 17.11 **Modern humans** *(Homo sapiens)* can be traced back about 100,000 years ago in Africa, from which birthplace they spread to West Asia. They reached East Asia and Australia 50,000 to 60,000 years ago and arrived in Europe only about 40,000 years ago. They were recent migrants into the Americas. (From *The Book of Life*, S.J. Gould, ed., New York:W.W. Norton & Co, 1995:232.)

mammoth, saber-toothed cat, and giant ox—in the upper Pleistocene. The large-scale annihilation of many game mammals has been called *Pleistocene overkill* by the evolutionist Paul S. Martin. Martin has noted that more than one-third of the genera of large mammals met extinction in Africa about 50,000 years ago and that nearly all of the larger game genera in North America perished 12,000 to 15,000 years ago. The latter event coincided with the migrations of early Mongoloid bands into North America via the Bering Strait. So devastating were the hunting activities of humans in North America that only a few forms of larger mammals survived, notably the bison and the pronghorn antelope. Such drastic reductions in game animals throughout the world may have been one of the precipitating factors in the transition some 10,000 years ago from a hunting economy to an agricultural economy.

ORIGIN OF MODERN HUMANS

The weight of evidence favors the thesis that modern humans were cradled in Africa. Most investigators would claim that modern humans with their distinctive features evolved in Africa between 150,000 and 100,000 years ago and displaced their predecessors as they migrated to various regions of the globe. Subsequently, regional ("racial") characteristics emerged in different geographical populations. This thesis is most often referred to as the *out-of-Africa theory*, although a variety of notations have proliferated—the "single-origin model," "monogenesis theory," "Noah's ark model," and the "replacement theory."

Africa was apparently the center from which at least two waves of migrations occurred: one took place about 1 million years ago by *Homo erectus* and the other about 150,000 years ago by *Homo sapiens*. In the first wave, the African *Homo*

Figure 17.12 Cro-Magnons, representatives of our own species, *Homo sapiens,* can be traced back about 40,000 years ago in Europe.
(Painting by Maurice Wilson; by permission of the Trustees of the British Museum—Natural History.)

erectus spread over Europe and Asia. In the second wave, modern *Homo sapiens* expanded from its continent of birth (Africa) and replaced *Homo erectus* and its descendants across the world.

There is a school of thought, chiefly identified with the paleoanthropologists Alan G. Thorne and Milford H. Wolpoff, that conceives of the modern human geographical populations as descending from different ancient hominid lineages evolving independently of one another. Thus, as seen in figure 17.13, the Indonesian *Homo erectus* ("Java man") was the early progenitor of the present native inhabitants of Australia, the Chinese *Homo erectus* ("Peking man") gave rise to modern Asians, the European *Homo erectus* gave rise to present-day Europeans, and the African *Homo erectus (Homo ergaster)* gave rise to present-day African populations. According to this view, gene flow between

the different geographical groups was sufficient to permit human populations to be maintained as one biological species. Simultaneously, the populations in different regions were sufficiently isolated to permit the development of distinct anatomic identities. This state of affairs is known as the *multiregional hypothesis* (also called "regional continuity hypothesis" or even the "candelabra hypothesis"). The Thorne-Wolpoff School would contend that geographic separation of the different early hominid branches did *not* lead to reproductive isolation, as might be expected of long-standing populations that are spatially separated and that differentiate along independent lines. It is, however, exceedingly difficult to imagine how several hominid geographical assemblages, diverging in different parts of the world, could evolve independently and yet repeatedly in the same direction leading only to one species, *Homo*

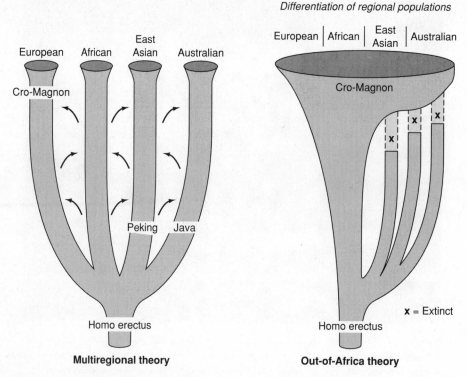

Figure 17.13 Origin of modern humans. The *multiregional theory* envisions a distant separation of the principal regional populations. Each regional population that stemmed from *Homo erectus* evolved independently and in parallel fashion over hundreds of thousands of years. Although each regional group evolved along its own distinctive direction, gene flow *(arrows represent gene exchange)* between neighboring geographical groups was sufficient to permit human populations to be maintained as one biological species. According to the *out-of-Africa theory,* regional differentiation occurred only after modern humans (Cro-Magnon) arose about 150,000 years ago in one place (Africa), and supplanted their predecessors as they migrated to various regions of the world.

Source: Modified from *Biology and Human Concerns,* Volpe, 4th ed., 1993:440.

sapiens. The parallel pattern of evolution associated with the absence of speciation is not hopelessly out of the question, but if it occurs, it must be the very rare exception to the normal process.

There is scarcely any disagreement that the world of humans today is a single large neighborhood. The once distinguishing features of geographical groups are becoming increasingly blurred by the interminglings and intermixings of peoples. Our present-day species lives in one great reproductive community.

DECIPHERING THE HUMAN GENOME

Genes responsible for human genetic disorders are being isolated at an astonishingly accelerated pace. Laboratories in many parts of the world are engaged in cooperative programs to determine the totality of genetic information encoded in the coiled DNA of the 46 human chromosomes. It is now feasible to map (locate) and sequence (arrange in order) the nucleotides of human DNA. An unprecedented international effort has been undertaken to create a "human gene dictionary." An estimated $200 million per year for the worldwide effort will be required to complete the project in 15 years. It is likely that the entire sequencing of nucleotides will be completed sooner than anticipated—by the end of the year 2003, which would represent the 50th anniversary of the announcement of the Watson-Crick model of DNA.

The genomic "book of life" is written in a language that has only four letters—A, C, G, T—representing the nucleic acid bases adenine, cytosine, guanine, and thymine. The DNA of the *haploid* human genome (single set of 23 chromosomes) contains approximately 3,000,000,000 (3×10^9) nucleotides. If, as shown in figure 18.1, a single page of the genomic book of life were to contain approximately 1,500 nucleotides, then the complete sequence of the haploid human genome would fill 2,000,000 pages. If contained in individual volumes, each approximating the length of this text ($\sim$ 300 pages), the genomic book of life would comprise an awesome stack of 6,667 volumes!

One of the goals of the Human Genome Project is to gain an appreciation of the manner in which variations in human DNA sequences cause genetic diseases, with the view of seeking improved therapies. The nucleotide sequence shown in figure 18.1 actually represents one of the first human genes isolated and sequenced—namely, the β-globin gene of the beta chain of hemoglobin. We shall meticulously dissect the components of this gene, which serves as a model for eukaryotic protein-coding genes. At the same time, we will examine how advanced knowledge of the molecular anatomy of the β-globin gene has permitted a deeper understanding of one of our best understood, yet most challenging genetic disorders—*thalassemia*.

The Array of Nucleotides in the Human β-globin Gene

```
CCCTGTGGAGCCACACCCTAGGGTTGGCCAATCTACTCCCAG
GAGCAGGGAGGGCAGGAGCCAGGGCTGGGCATAAAAGTCAG
GGCAGAGCCATCTATTGCTTACATTTGCTTCTGACACAACTGT
GTTCACTAGCAACCTCAAACAGACAGGATGGTGCACCTGACT
CCTGAGGAGAAGTTGGCCTTACTCGCCCTGTGGGGCAAGGTG
AACGTGGATGAAGTTGGTGGTGAGGCCCTGGGCAGGTTGGTA
TCAAGGTTACAAGACAGGTTTAAGGAGACCAATAGAAACTGG
GCATGTGGAGACAGAGAAGACTCTTGGGTTTCTGATAGGCAC
TGACTCTCTCTGCCTATTGGTCTATTTTCCCACCCTTAGGCTG
CTGGTGGTCTACCCTTGGACCCAGAGGTTCTTTGAGTTCTTTG
GGGATCTGTCCACTCCTGATGCTGTTATGGGCAACCCTAAGG
TGAAGGCTCATGGCAAGAAAGTGCTCGGTGCCTTTAGTGATGG
CCTGGCCCACCTGGACAACCTCAAGGGCACCTTTGCCACACT
GAGTGAGCTGCACTGTGACAAGCTGCACGTGGATCCTGAGAA
CTTCAGGGTGAGTCTATGGGACCCTTGATGTTTTCTTTCCCCT
TCTTTTCTATGGTTAAGTTCATGTCATAGGAAGGGGAGAAGTA
ACAGGGTACAGTTTAGAATGGGAAACAGACGAATGATTGCAT
CAGTGTGGAAGTCTCAGGATCGTTTTAGTTTCATTTATTTGCT
GTTCATAACAATTGTTTTCTTTTGTTTAATTCTTGCTTTCTTTT
TTTTCTTCTCCGCAATTTTTACTATTATACTTAATGCCTTAA
CATTGTGTATAACAAAAGGAAATATCTCTGAGATACATTAAG
TAACTTAAAAAAAAACTTTACACAGTCTGCCTAGTACATTAC
TATTTGGAATATATGTGTGCTTATTGCATATTCATAATCTCC
CTACTTTATTTTCTTTTATTTTTAATTGATACATAATCATTATA
CATATTTATGGGTTAAAGTGTAATGTTTTAATATGTGTACACA
TATTGACCAAATCAGGGTAATTTTGCATTTGTAATTTTAAAAA
ATGCTTTCTTCTTTTAATATACTTTTTTGTTTATTTTATTTCTA
ATACTTTCCCTAATCTCTTTCTTTCAGGGCAATAATGATACAA
TGTATCATGCCTCTTTGCACCATTCTAAAGAATAACAGTGATA
ATTTCTGGGTTAAGGCAATAGCAATATTTCTGCATATAAATA
TTTCTGCATATAAATTGTAACTGATGTAAGAGGTTTCATATTG
CTAATAGCAGCTACAATCCAGCTACCATTCTGCATTTATTTTA
TGGTTGGGATAAGGCTGGATTATTCTGAGTCCAAGCTAGGCC
CTTTTGCTAATCATGTTCATACCTCTTATCTTCCTCCCACAGC
CCCTGGGCAACGTGCTGGTCTGTGTGCTGGCCCATCACTTTGG
CAAAGAATTCACCCCACCAGTGCAGGCTGCCTATCAGAAAGT
GGTGGCTGGTGTGGCTAATGCCCTGGCCCACAAGTATCACTA
AGCTCGCTTTCTTGCTGTCCAATTTCTATTAAAGGTTCCTTTG
TTCCCTAAGTCCAACTACTAAACTGGGGGATATTATGAAGGG
CCTTGTGCATCTGGATTCTGCCTAATAAAAAACATTTATTTC
ATTGC
```

Figure 18.1 The array of nucleotides in the human β-globin gene. The sequences of nucleotides are deciphered in Figure 18.4.

(Based on raw data from a study by Richard M. Lawn and his colleagues at the California Institute of Technology, 1980.)

EXON/INTRON STRUCTURE OF GENES

The earlier story of gene action (see chapter 8), based on the fine investigations by Linus Pauling and Vernon Ingram, was elegantly simple. However, the structure and expression of genes are much more intricate than initially envisioned. Up until the 1970s, the protein-coding gene was viewed as an uninterrupted, continuous coding stretch of DNA. In the late 1970s, it became clear that the eukaryotic gene contains unexpressed sequences that interrupt the continuity of the genetic information. The coding sequences have been termed *exons,* whereas the noncoding intervening sequences have been called *introns.* The entire gene is transcribed as a long RNA precursor, commonly referred to as the *primary RNA transcript.* The introns of the primary RNA transcript are then clipped out as the exons are spliced together to yield the intact coding sequence in the functional messenger RNA (fig. 18.2). Thus, the final form of messenger RNA (mRNA) has been shortened appreciably by the removal of "useless" intervening sequences. Accurate "cut and paste" is assured by specific enzymes that recognize precise signals at splice junctions in the primary transcript. There is no rule that governs the number of introns. The gene for the beta chain of human hemoglobin contains two introns, whereas the variant gene that causes Duchenne muscular dystrophy has more than 60 introns. Interestingly, nearly all bacteria and viruses have streamlined their structural genes so as to contain no introns whatsoever. It is likely that strong selection pressures were exerted on these microorganisms in order to retain a small genome size.

Exons apparently partition proteins into meaningful subdivisions, or *domains,* that perform specific functions. From an evolutionary point of view, the separately coded protein domains might be an efficient mechanism for the reassortment of different functional domains into new proteins. Stated another way, introns may have played a major role in the structuring of genes by moving exons around and inserting them in other genes.

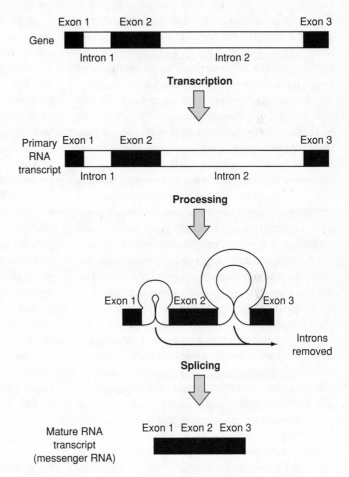

Figure 18.2   The primary transcript of the gene for the β chain of hemoglobin contains two noncoding intervening sequences (introns), which are removed. The coding sequences (exons) are spliced together to yield the intact message in the mature messenger RNA. Additional detailed information is provided in figure 18.3.

(Modified from Volpe, E.P. *Biology and Human Concerns,* 4th ed., 1993, Dubuque IA:WCB.)

This mechanism was termed *exon shuffling* by Walter Gilbert of Harvard University in the 1970s. One need only examine the exon/intron structure of the human gene that encodes the membrane receptor for low-density lipoprotein (LDL) to discern Gilbert's hypothesis at work. This *LDL receptor* gene has 18 exons that encode six protein domains in the receptor. The several exons of one domain have an inordinate sequence similarity to regions of the gene for *epithelial growth factor.* The exons of another domain are homologous to a gene for a particular blood protein *(C9 complement factor).* It is as if the mosaic of domains for LDL receptor were assembled by the shuffling of genes from different transcription units.

In the 1940s, Nobel laureates G.W. Beadle and Edward L. Tatum, then at the California Institute of Technology, set forth with special clarity and thoroughness the *one gene-one enzyme concept,* which states that each gene controls the formation of a single enzyme and thereby a single metabolic reaction. With increasing knowledge of the nature of genes in recent years, the one gene-one enzyme thesis has been greatly reinforced and expanded. It has been extended to include all cellular proteins, not necessarily those that are enzymes, and has been more precisely defined as the *one gene-one polypeptide principle.* But now this famous dictum has been flatly contradicted by the finding that one structural gene can code for *two or more different polypeptides.* A striking example is the gene that codes for *fibronectin,* an adhesive glycoprotein that enables cells to adhere to the extracellular matrix. Fibronectin is a multifunctional protein that is involved in many processes—embryogenesis (cell migration), hemostasis, wound healing, and connective tissue integrity. The variety of forms of fibronectin is encoded by a single large gene. Transcription produces a single large primary RNA transcript, which gives rise to *as many as 20 different messenger RNAs* through a complicated pattern of alternative splicing. Removal of different combinations of introns and exons gives rise to proteins with different functions.

REPETITIVE DNA

In the late 1960s, Roy J. Britten, then at the Carnegie Institution of Washington in Washington, DC, and Eric H. Davidson, of the California Institute of Technology, startled the genetic community with the announcement that the DNA of eukaryotes is *repetitive*—that is, there are many copies of seemingly nonessential DNA sequences (of various lengths and compositions) that do *not* code for protein. The proportion of the genome taken up by repetitive sequences varies widely among taxa. In yeast, this proportion amounts to about 20 percent; in mammals, up to 60 percent of the DNA is repetitive. In the human haploid genome, an estimated 2 million (2×10^9) nucleotides have no coding functions, or are not transcribed.

Some of the repetitive fractions are made up of short sequences, from a few to hundreds of nucleotides long, which are repeated on the average 500,000 times. Other repetitive fractions consist of much longer sequences, thousands of nucleotides on the average, which appear in the genome up to hundreds of times. The repetitive sequences are *similar but not identical* among individuals, inasmuch as they represent a great reservoir for mutational changes. Indeed, the genetic variations within the repetitive DNA stretches are so immense that each single human can be distinguished by a DNA "fingerprint" based on the heritable variations in the repeated sequences. Elaborate DNA fingerprinting techniques are employed to reveal an individual's genetic uniqueness.

One of the more interesting classes of repetitive sequences in mammals is the *Alu* elements, which tend to be evenly interspersed throughout the genome. The name derives from the fact that many of these repetitious sequences in humans contain recognition sites for the restriction enzyme *Alu* I. The entire group has been referred to as the *Alu* family. The *Alu* sequences are 200 to 300 nucleotides in length, of which there are an estimated one million copies in the human genome.

They constitute between 5 and 10 percent of the total (diploid) human genome. Various debatable roles have been ascribed to the *Alu* elements, from "molecular parasites" to initiation sites of DNA synthesis. Perhaps the most important consideration is that *Alu* sequences are transposable. The majority of *Alu* sequences are flanked on either side by repeat sequences similar to those of bacterial insertion sequences (see chapter 7). The potential mobility of such a large number of short nucleotide sequences inspires solemn awe.

It is still a subject of great interest that most of the DNA sequences in higher organisms are not transcribed at all. Selectionists have argued that the extra, noncoding DNA must have some important function or it would not be there. Accordingly, functions of presumed benefit have been suggested. The numerous sequences of noncoding DNA are said to be present because they contribute to chromosome structure, or because they are required for regulating trancription or translation, or because over the long term they facilitate the recombination of DNA to provide the raw material for new coding genes.

Several investigators, including Nobel laureate F. H. C. Crick, assert that much of the extra DNA is "junk." In other words, the repetitious DNA has little specificity and conveys no selective advantage to the organism. The only function of the extra DNA sequences is self-survival or self-preservation. The junk DNA spreads through the cell's genome in much the same manner as a not-too-harmful parasite spreads in its host. The noncoding sequences constantly endeavor to expand to the limit tolerated by the cell's genome. They are the ultimate parasite, or *selfish DNA*.

Amidst this sea of repetitious DNA of uncertain function lie islands of *unique-sequence DNA* responsible for the coding of proteins. These protein-coding genes are essentially the familiar "structural" genes of classical genetics. The diploid human genome contains about 6.0×10^9 base pairs of DNA and over 300 times more unique-sequence DNA than in the genome of the bacterium *E. Coli*. Thus, the human

genome has the potential of encoding an enormous amount of genetic information. Victor McKusick of the Johns Hopkins School of Medicine has listed at least 7,000 variant phenotypes associated with unique-sequence genes. The list of variant genes is updated annually, and since most are responsible for specific diseases, McKusick whimsically refers to his collection of mutant genes as the "Morbid Anatomy of the Human Genome." At the completion of the Human Gene Project, it is anticipated that at least 100,000 protein-coding genes will have been mapped and sequenced.

MICROANATOMY OF THE β-GLOBIN GENE

In our consideration of the β-globin gene, we will present the entire scheme related to the production of the beta-chain polypeptide—that is, all the structural and regulatory factors that are associated with the production of the polypeptide. As currently understood, the expression of a protein-coding gene is regulated by several kinds of DNA sequences, each of which binds specific proteins that determine the rate and quality of transcription of the protein-coding gene. Figure 18.3 provides an initial glimpse of the complexity of the gene, with its flanking regions.

The β-globin gene is a relatively small gene, consisting of three exons and two introns. The exons contain codons that prescribe a 146-amino acid polypeptide chain: Exon 1 codes for amino acid residues 1-30; Exon 2, residues 31-104; and Exon 3, residues 105-146. Important roles in gene function are played by three flanking regions—a promoter region, a 5′ leader, and a 3′ trailer—all three of which are *noncoding* segments

Two well-characterized promoters, which reside upstream of the gene, regulate gene expression. One of the promoters, the TATA box, lies 25 to 30 nucleotides upstream from the initiation site at which transcription begins. This short sequence serves to bind the enzyme (RNA polymerase) that

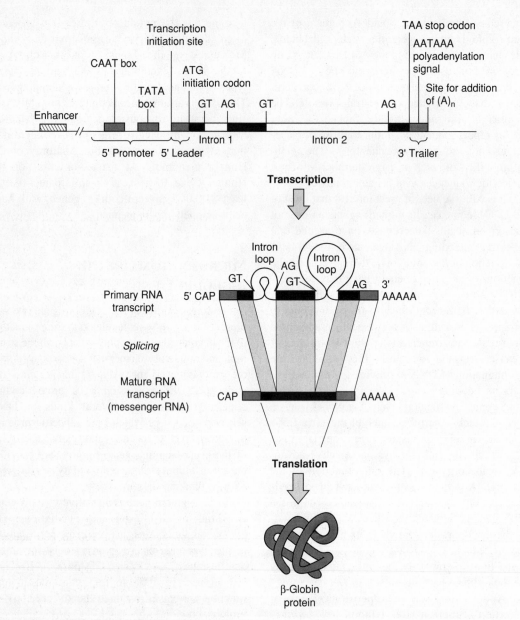

Figure 18.3 Structure and expression of the human β-globin gene, including the three flanking regions—a promoter region, a 5′ leader, and a 3′ trailer. See text for details.

synthesizes the RNA chain. The binding of the polymerase can occur only if an appropriate series of transcription factors (TFIIB, TFIID, among others) have previously bound to the TATA box. Another upstream promoter element occurs 70 to 90 nucleotides from the initiation site, typically involving the CAAT box, which is also concerned with the regulation of RNA transcription by acting as a binding site for regulatory signals for the RNA polymerase. The term "promoter" is reserved for elements that have relatively fixed spatial relationships with the genes they control. Reading of the gene—that is, its production of mRNA—is also facilitated by the presence of a distantly located *enhancer.* Enhancers differ from promoter elements in that they can be inverted without diminishing their effects and can operate at considerable distances (upstream or downstream) from the genes they control. Enhancers facilitate the assembly and activation of the transcriptional complex at the promoter site.

INTERPRETING THE LINEAR ARRAY OF NUCLEOTIDES

In the 1980s, a team of investigators led by Richard M. Lawn of the California Institute of Technology reported the complete nucleotide sequence of the human β-globin gene. Figure 18.4 represents an interpretation of the raw data, shown in figure 18.1 as a linear array of nucleotides. As is the convention in molecular genetics, figure 18.4 depicts the nucleotide sequence of the DNA strand which is *complementary* to the DNA strand (transcriptional template) that codes for the messenger RNA strand. Hence, it should not be surprising to see methionine (the familiar initiation codon) represented as ATG.

The TATA box is shown in the first line of the promoter region in figure 18.4. As is the case with all consensus sequences, C's and G's occasionally creep in. The TATA box of the β-globin gene starts with a C, and is represented by CATAAAA. Further upstream, at approximate position –75, the CAAT box is represented by CCAAT. Mutational changes (such as deletions) of nucleotides in the TATA box as well as the CAAT box result in profound changes in the efficiency of transcription.

An interesting feature is the presence of a 5′ *leader* at the leading end of the gene (figure 18.4). Although these leading sequences *are included* in the mRNA, they do *not* code for amino acids. The 5′ leader may function in a manner similar to the

leader of a movie film. The 5′ leader terminates in the initiation triplet ATG, which codes for methionine. This amino acid almost invariably functions as the starting codon. Since mature proteins rarely begin with methionine, the initiating methionine is typically cleaved from most proteins. This is the case with the β-globin polypeptide; the initiating methionine is cleaved off and the first amino acid of the mature polypeptide is valine.

For exon 1 (as well as the other exons), the amino acid for each codon is written above the genomic sequence in figure 18.4. The intron sequences between exon 1 and 2 and between exon 2 and 3 are initially transcribed into RNA but are then sliced out before the final messenger RNA molecule is exported to the cytoplasm. Certain sequences around the exon/intron junctions serve as recognition sites for splicing enzymes—typically, an intron begins with GT (the 5′ donor site) and ends with AG (the 3′ acceptor site). In figure 18.4, the GT and AG dinucleotides at the beginning and end of the introns are underlined. Each of the exon/intron boundaries are potential sites of mutations that affect gene expression.

The nascent RNA strand is synthesized in an upstream (5′) to downstream (3′) direction at about 40 nucleotides per second until the transcription terminator is reached. In the β-globin gene, transcription ends with the terminator codon TAA. A methylguanylate residue ("cap") is added to the 5′ end of the primary RNA transcript. At the 3′ end, a poly(A) tail (an unbroken stretch of 100 to 200 adenine nucleotides) is added to aid in the transport of the mature mRNA from the nucleus to the cytoplasm. This poly(A) tail is not specified by genomic DNA but is added after transcription by a poly(A) polymerase in response to the sequence AATAAA in the 3′ noncoding trailer region. The probable role of the poly(A) tail is to protect the vulnerable mRNA from degradation by ribonucleases in the cytoplasm. Inasmuch as cytoplasmic ribonucleases eventually digest the tail, it is likely that the life expectancy of the mRNA is set by the tail. The poly(A) tail has been likened to a slow-burning fuse.

ccctggagccacaccctagggttggccaatctactcccaggagcagggagggcaggagccagggctgggcataaaa
gtcagggcagagccatctattgctt **Promoter region**

ACATTTGCTTCTGACACAACTGTGTTCACTAGCAACCTCAAACAGACACCATG **5′ leader**

Val His Leu Thr Pro Glu Glu Lys Ser Ala Val Thr Ala Leu Trp Gly Lys Val Asn Val
GTG CAC CTG ACT CCT GAG GAG AAG TTG GCC TTA CTC GCC CTG TGG GGC AAG GTG AAC GTG

Asp Glu Val Gly Gly Glu Ala Leu Gly Arg
GAT AAG GTT GGT GGT GAG GCC CTG GGC AGG **Exon 1**

GTGGTATCAAGGTTACAAGACAGGTTTAAGGAGACCAATAGAAACTGGGCATGTGGAGACAGAGAAG
ACTCTTGGGTTCTGATAGGCACTGACTCTCTCTGCCTATTGGTCTATTTTCCCACCCTTAG **Intron 1**

Leu Leu Val Val Tyr Pro Trp Thr Gln Arg Phe Phe Glu Ser Phe Gly Asp Leu Ser Thr Pro
CTG CTG GTG GTC TAC CCT TGG ACC CAG AGG TTC TTT GAG TTC TTT GGG GAT CTG TCC ACT CCT

Asp Ala Val Met Gly Asn Pro Lys Val Lys Ala His Gly Lys Lys Val Leu Gly Ala Phe
GAT GCT GTT ATG GGC AAC CCT AAG GTG AAG GCT CAT GGC AAG AAA GTG GTC GGT GCC TTT

Ser Asp Gly Leu Ala His Leu Asp Asn Leu Lys Gly Thr Phe Ala Thr Leu Ser Glu Leu
AGT GAT GGC CTG GCC CAC CTG GAC AAC CTC AAG GGC ACC TTT GCC ACA CTG AGT GAG CTG

His Cys Asp Lys Leu His Val Asp Pro Glu Asn Phe Arg
CAC TGT GAC AAG CTG CAC GTG GAT CCT GAG AAC TTC AGG **Exon 2**

GTGAGTCTATGGGACCCTTGATGTTTTCTTTCCCCTTCTTTTCTATGGTTAAGTTCATGTCATAGGAAGGGGAGAAGTAACAGGGTA
CAGTTTAGAATGGGAAACAGACGAATGATTGCATCAGTGTGGAAGTCTCAGGATCGTTTTAGTTTCATTTATTTGCTGTTCATA
ACAATTGTTTTCTTTTGTTTAATTCTTGCTTTCTTTTTTTTTCTTCTCCGCAATTTTTACTATTATACTTAATGCCTTAA
CATTGTGTATAACAAAAGGAAATAGTCTCTGAGATACATTAAGTAACTTAAAAAAAAAACTTTACACAGTCTGCCTAGTACATTA
CTATTTGGAATATATGTGTGCTTATTTGCATATTCATAATCTCCCTACTTTATTTTCTTTTATTTTAATTGATACATAATCATTTAT
ACATATTTATGGGTTAAAGTCGTAATGTTTTAATATGTGTACACATATTGACCAAATCAGGGTAATTTTCAGACTTTGTAATTTTAAA
AAATGCTTTCTTCTTTTAATATACTTTTTTGTTTATTTTATTTCTAATACTTTCCCTAATCTCTTTCTTTCAGGGCAATAATGATA
CAATGTATCATGCCTCTTTGCACCATTCTAAAGAATAACAGTGATAATTTCTGGGTTAAGGCAATAGCAATATTTCTGCATATA
AATATTTCTGCATATAAATTGTAACTGATGTAAGAGGTTTCATATTGCTAATAGCAGCTACAATCCAGCTACCATTCTGCATTT
ATTTTATGGTTGGGATAAGGCTGGATTATTCTGAGTCCAAGCTAGGCCCTTTTGCTAATCATGTTCATACCTCTTATCTTCCTCC
CACAG **Intron 2**

Leu Leu Gly Asn Val Leu Val Cys Val Leu Ala His His Pha Gly Lys Glu Phe Thr
CCC CTG GGC AAC GTG CTG GTC TGT GTG CTG GCC CAT CAC TTT GGC AAA GAA TTC ACC

Pro Pro Val Gln Ala Alla Tyr Gln Lys Val Val Ala Gly Val Ala Asn Ala Leu Ala
CCA CCA GTG CAG GCT GCC TAT CAG AAA GTG GTG GCT GGT GTG GCT AAT GCC CTG GCC

His Lys Tyr His
CAC AAG TAT CAC **Exon 3**

TAAGCTCGCTTTCTTGCTGTCCAATTTCTATTAAAGGTTCCTTTGTTCCCTAAGTCCAACTAC
TAAACTGGGGGGATATTATGAAGGGCCTTGTGCATCTGGATTCTGCCTAATAAAAAACATTTATTTTCATTGC **3′ trailer**

Figure 18.4 The organization of the normal human β-globin gene and its flanking sequences. The nucleotide sequences of the anticoding strand of DNA are displayed from the upstream (5′) to the downstream (3′) direction. Capital letters of the exons are highlighted; the amino acid sequences of the β-globin polypeptide are shown on the line above the triplets of nucleotides. The upstream promoter region is represented by lowercase letters; the CAT and TATA boxes are shaded. The ATG initiator and TAA terminator are both shaded and underlined. The GT and AG dinucleotides at the beginning and end of the introns are underlined.

(Based on studies by Richard M. Lawn and his colleagues at the California Institute of Technology, 1980.)

THALASSEMIA AND MUTATIONAL CHANGES

Thalassemia is a serious, potentially fatal, blood disorder that is associated with the marked suppression of hemoglobin production. Once thought to be a single disease entity, thalassemia is now known to be a diverse group of hemolytic anemias, characterized by the absence or decreased synthesis of one of the normal globin chains of hemoglobin. Our attention is focused here on β-thalassemia, in which a molecular defect in the *β*-globin gene prevents the normal expression of the gene. As we learned earlier, the *β*-globin gene is not imperative in fetal life and is not even fully expressed until after birth. Hence, β-thalassemia in its full expression is a crippling disease of early childhood.

Beta chain production is absent or inadequate to meet physiologic demands in patients homozygous for the defective *β*-globin gene. Affected children are acutely anemic. Newly produced blood cells are so altered that most fail to leave the bone marrow. Those that manage to enter the circulation are small, pale, and have a shortened life span. Tissues are deprived of oxygen as the bone marrow fails to maintain an effective hemoglobin level in the body. Recurrent infections occur and cardiac failure ensues. If untreated, 80 percent of affected children will die during the first year of life.

Heterozygous individuals tend to be asymptomatic. Because individuals who are heterozygous for β-thalassemia are somewhat resistant to malaria, the small reproductive edge they enjoy has caused this disease to become endemic around the Mediterranean and Black Sea, where malaria was prevalent. In people of Mediterranean, Near and Middle Eastern, and Southeast Asian origins, frequencies of variant *β*-globin genes range from 5 percent to 20 percent. β-thalassemia is but another striking example of balanced polymorphism (see chapter 8).

The level of *β*-chain gene function is governed by a *variety of different mutational changes within and surrounding the β-globin gene.* Some patients with β-thalassemia are unable to synthesize any beta chains whatsoever; others experience reduc-

tions in the synthesis of the beta chains. The molecular lesions underlying the β-thalasssemias are highly varied. At least 150 different specific mutations leading to β-thalassemia have been defined, affecting several aspects of gene function: gene transcription, RNA processing, and RNA translation. Nucleotide substitutions occur in the promoter region, but these mutations often do not result in a completely inactive *β*-globin gene. Rather, binding is reduced and the gene continues to produce mRNA at a reduced rate. Mutations can occur that affect the GT or AG dinucleotides at the intron junctions, resulting in incorrect splicing and an abnormal mRNA.

Another type of error involves the addition or deletion of a single nucleotide within one of the exons, which typically results in an altered sequence of amino acids in the polypeptide (fig. 18.5). For example, if the base sequence ATGCTTCTC is normally read [ATG][CTT][CTC], then an insertion of the base C between [ATG] and [CTT] would lead to a shift in the "reading frame," as follows: [ATG][CCT][TCT][C..]. Such modifications are called *frameshift* mutations (fig.18.5). Another type of error is a *nonsense* mutation, in which a codon that would ordinarily be decoded into an amino acid is altered to one that terminates translation.

SIGNIFICANCE OF THALASSEMIA

Investigators were unprepared to find the prodigious variety of different mutations that can lead to reduction of the output of the *β*-globin gene and cause β-thalassemia. Different geographical populations have their own predominant mutations. Much of the same diversity of mutations is being revealed in other human genes as they are isolated and studied, such as cystic fibrosis, Duchenne muscular dystrophy, and phenylketonuria. Today we are inclined to view sickle-cell anemia as rather unusual. Although sickle-cell anemia occurs in many different regions, it is invariably produced by the identical genetic error—a single base change that results in the substitution of valine for glutamic acid at the same precise point in the β-chain in every affected individual.

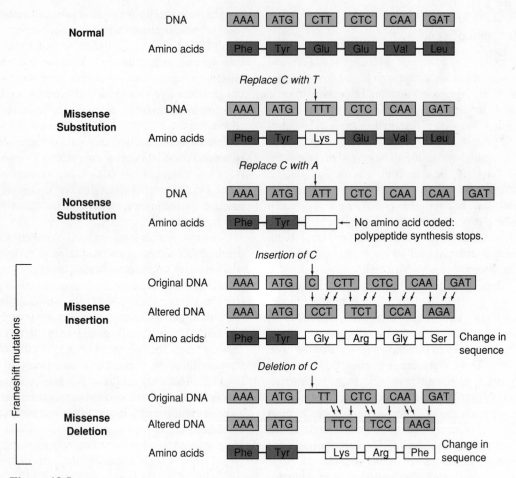

Figure 18.5 **The molecular basis of point mutation.** Each substitution, deletion, or insertion is a mutational event that alters the DNA sequence and, in turn, the polypeptide (protein) product.

During the past two decades, medical care of thalassemia has been improved by our expanded knowledge of the molecular basis of the disease. We may recall that fetal cells have a relatively high affinity for oxygen, thereby facilitating the transport of oxygen across the placenta. Shortly before birth, the β-globin gene is turned on and the γ-globin gene is turned off. As a result, in normal children over six months of age, hemoglobin A makes up about 96 percent of the total hemoglobin and fetal hemoglobin makes up less than one percent, this largely sequestered in the bone marrow.

Armed with modern molecular information, medical scientists have successfully reversed the ontogenetic process by turning the γ-globin gene on again. At present, a pharmacologic approach is being used; infusions of the chemical agent, arginine butyrate, into patients have resulted in the stimulation of γ-globin messenger RNA and sustained levels of fetal hemoglobin. The ultimate treatment is *gene therapy:* the delivery of functional globin genes to hematopoietic stem cells. As we await gene therapy (see chapter 21), bone marrow transplantation is now in hand for thalassemic patients. Prenatal detection of β-thalassemia is currently available. Screening and genetic counseling programs have allowed informed choice for couples at risk.

GENOMES OF CHIMPANZEES AND HUMANS

Chimpanzees *(Pan troglodytes)* and humans *(Homo sapiens)* are two of the most thoroughly studied species, at both the organismal and molecular levels. In recent years, a variety of protein molecules of the two species have been analyzed by amino acid sequencing, immunology, and electrophoresis. All the molecular data agree in showing that the chimpanzees and humans are remarkably similar. The two species have completely identical fibrinopeptides, cytochrome *c,* and alpha, beta, and gamma chains of hemoglobin. With respect to myoglobin and the delta chain of hemoglobin, only a single acid replacement separates the human polypeptide chain from that of the chimpanzee. Based on protein analyses, the sequences of human and chimpanzee polypeptides are, on the average, more than 99 percent identical. Stated another way, the average human polypeptide is *less than 1 percent* different from its chimpanzee counterpart. The information on proteins has been reinforced by nucleic acid studies, which have revealed that our nuclear DNA is 98 to 99 percent identical to that of the chimpanzee. It would seem that the exceedingly meager molecular differences, both at the protein and DNA levels, are much too small to account for the substantial anatomical and behavioral differences between the two species.

It has been suggested that the major organismal differences between the two species are largely based on changes in the *regulatory sequences* that control the expression of structural (protein-coding) genes rather than on changes in the structural genes themselves. In particular, the complex biological differences between humans and chimpanzees may stem primarily from mutational changes in regulatory units, such as promoters, enhancers, and transcription factors. A subtle change in a regulatory sequence that codes for a particular transcription factor might result, for example, in the prolongation of the growth period of the brain of the human fetus, thereby permitting additional time for the development of greater complexity of the brain.

Gross chromosomal rearrangements (inversions and translocations) may play a prominent role in shifting a given regulatory gene from its normal position to a new location in another chromosome. Despite appreciable homology of the two chromosomal sets, the banding patterns of certain chromosomes of humans and chimpanzees are clearly different. Specifically, chromosomes 4, 9, and 12 of the two species show evidence that segments have been reshuffled. Changes in gene regulation brought about by the remodeling of chromosomes could be very profound. In essence, regulatory genes determine when, and the extent to which, the production of structural proteins will be enhanced or repressed. One could say that *when* a protein is produced *quantitatively* is at least as important evolutionarily as *what* protein is produced *qualitatively.*

19

MICROEVOLUTION, MACROEVOLUTION, AND PUNCTUATED EQUILIBRIA

Evolutionary changes in populations are ordinarily visualized as gradual, built upon many small genetic variations that arise and are passed on from generation to generation. The shifting gene frequencies in local populations may be thought of as *microevolution*. The progressive replacement of light-colored moths by dark moths in industrial regions of England exemplifies the microevolutionary processes. Most population geneticists subscribe to the view that the same microevolutionary processes have been involved in the major transformations of organisms over long spans of geologic time *(macroevolution)*. The traditional outlook is that small variations gradually accumulate in evolving lineages over periods of millions of years. If we were to recover a complete set of fossil specimens of a lineage, we would expect to find a graded series of forms changing continuously from the antecedent species to the descendant species. The temporally intermediate forms would be intermediate morphologically. Under this perspective, our inability to find transitional fossil forms between the ancestral and descendant populations would represent merely the imperfect nature of the fossil record.

For several years, Stephen Jay Gould of Harvard University and Niles Eldredge of the American Museum of Natural History have questioned the conventional view that evolutionary changes in the distant past are principally the outcome of the gradual accumulation of slight inherited variations. They advocate that most evolutionary changes have consisted of rapid bursts of speciation alternating with long periods in which the individual species remain virtually unmodified. Gould and Eldredge maintain that most lineages display such limited morphological changes for long intervals of geologic time as to remain in stasis, or in "equilibria." Conspicuous or prominent evolutionary changes are concentrated in those brief periods ("punctuations") when the lineages actually split or branch. This is the now celebrated hypothesis of *punctuated equilibria*.

RAPID EVOLUTION IN ANIMAL SPECIES

In the discussion of geographic speciation in chapter 10, there was scarcely any hint that a new animal species could arise abruptly. Species formation was held to occur by the gradual genetic divergence of geographically separated (allopatric) populations. In the 1970s, Hampton L. Carson of the University of Hawaii in Honolulu demonstrated that animal speciation can occur rapidly, within relatively few generations. This has been evidenced in populations of flies of the family Drosophilidae, commonly called the "picture-winged" flies. These extraordinary flies are indigenous to the high-altitude forests of the Hawaiian Islands (Archipelago) in the tropical region of the Pacific Ocean. The geologically very recent Hawaiian Islands, created by volcanic action, are the most isolated ocean islands in the world, at least 2,000 miles from any continent. The youngest and largest island in this archipelago is Hawaii, which has received much of its unique fauna from the adjacent island of Maui.

Chance has played the major role in the establishment of the interesting array of species of picture-winged flies on Hawaii. There are 26 species of pictured-winged flies inhabiting an island that is less than 700,000 years old. On the basis of analysis of chromosomal inversions, Carson infers that each species on Hawaii was descended from a few waif ancestors (indeed, possibly a single gravid female) that arrived fortuitously from Maui and other older islands. After each successful colonizing event, morphologically and genetically distinct species have been formed. An important aspect is that the migrant populations gave rise to new species rather than simply to new colonies of the parental species.

The process of almost instantaneous speciation seems to have occurred repeatedly in the proliferation of Hawaiian flies. Each small isolate of the ancestral population contained only a small fraction of the ancestral gene pool. From such a disrupted or reduced gene pool, natural selection rebuilt and reorganized a genetic system that contained unique elements and combinations. New genotypes arose that would not have appeared or would not have survived in the parental populations. Reproductive isolation most likely developed largely as an incidental by-product of the new gene complexes. It is striking that the evolutionary events that produced

the varied species of picture-winged flies on Hawaii consumed less than 700,000 years.

THE FOUNDER PRINCIPLE AND RAPID DIVERSITY

It may be recalled (chapter 9) that the Founder principle is a special case of Sewall Wright's genetic drift. The Founder principle is the establishment of a new population by a few migrant individuals who carry only a small fraction of the total genetic variability of the parental population. As first enunciated by the Harvard evolutionist Ernst Mayr in 1956, such small, peripherally isolated populations may undergo rapid evolutionary divergence to form new species. In fact, such rapidly evolving, locally isolated populations may have been the source of many evolutionary novelties in the distant past. The genetic restructuring of a Founder population is likely to be novel, since strong selection pressures are operating on an altered gene pool in an alien habitat. Morphological differentiation often occurs in conjunction with the evolution of reproductive barriers separating the colonizing population from the parent population. Mayr's scheme of the rapid emergence of a new species by the Founder principle provides one of the essential ingredients in the punctuational mode of evolution expounded by Gould and Eldredge.

TENET OF PUNCTUATED EQUILIBRIA

A single line of descent, or lineage, may persist for long reaches of geologic time. As small changes accumulate over periods of millions of years within one lineage, the descendant populations may eventually be recognized as a species distinct from the antecedent populations. The persistent accumulation of small changes within a lineage has been termed *phyletic gradualism*, and the transformation of a lineage over time has been termed *anagenesis*. As depicted in figure 19.1, a new species (labeled "B") arises from the slow and steady transformation of a large antecedent population ("A"). If only transformation occurred, then life would cease as single lineages became extinct. Hence, a new species ("C" in fig. 19.1) can also arise by the splitting or branching of a lineage. The splitting of one phyletic lineage into two or more lines is termed *cladogenesis*. But, here again, the splitting is conceived of as proceeding slowly and gradually, with the two branching lineages progressively diverging, without significant reduction in population size. Thus, paleontologists tend to view lineage splitting in terms of gradual morphological divergence. Viewed as a slow process, lineage splitting becomes reduced to a special case of the phyletic model.

Gould and Eldredge maintain that phyletic gradualism is much too slow to produce the major

events of evolution. In fact, the morphological changes in successive populations of lineage are usually directionless and only minor temporal variations occur. The theory of punctuated equilibria advocates that the prominent episodes of evolution in the history of life are associated with the splitting of lineages, but the splitting is not seen as slow and steady cladogenesis. The new species arises through rapid evolution when a small local population becomes isolated at the margin of the geographic range of the parent species (fig. 19.2). Indeed, the successful branching of a small isolated population from the periphery of the parental range virtually assures the rapid origin of a new species. In geologic time, the branching is sudden—thousands of

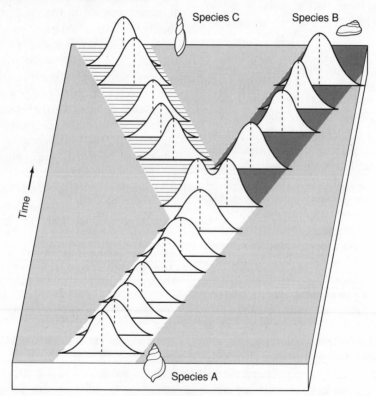

Species C   Species B

Time

Species A

Phyletic Gradualism

Figure 19.1 According to phyletic gradualism, a lineage consists of a graded series of intermediate forms connecting ancestral and descendant organisms. Species A gradually transforms into species B and may split into a slowly evolving species C.

Redrawn by Phil Mattes, in Volpe, E.P., *Biology and Human Concerns,* 4th ed., 1993, Dubuque IA: WCB. fig. 41.1, p. 443.

years or less, compared to the millions of years of longevity of the species itself.

The punctuational change does not entail any unconventional evolutionary phenomenon. The sudden appearance of a species is fully consistent with Ernst Mayr's rapid selection in peripheral isolates. The morphological differentiation of geographical isolates occurs early and rapidly, especially when the population is small and adapting to the new atypical environment at the periphery of the range. The expectation is that once the new species is established little evolutionary

change will occur in that species over geologic time. The long stasis of fossil species is disconcerting to evolutionists who endorse the slow phyletic transformation of species.

Given the thesis of punctuated equilibria, the fossil record is a faithful rendering of the evolutionary processes. If a species remains essentially static during its long lineage, then one would not anticipate finding a continuous series of transitional fossils. Moreover, if a descendant species arises by allopatric speciation in small, peripheral populations, then it would be rare to find evidence of

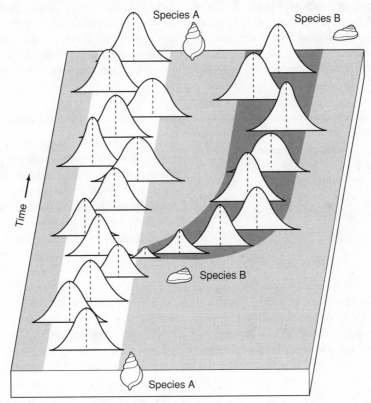

Punctuated Equilibria

Figure 19.2 The theory of punctuated equilibria proposes that a small isolate from the parental population (species A) evolves rapidly into a new entity (species B), which undergoes a long period of stasis during which little or no morphological change takes place. Species B does undergo minor structural modifications with time, but these modifications represent merely oscillations around the mean.

Redrawn by Phil Mattes, in Volpe, E.P., *Biology and Human Concerns,* 4th ed., 1993, Dubuque IA: WCB. fig. 41.2, p. 443.

preservation of a small population evolving in a relatively brief period of time. Thus, the so-called "gaps" in the fossil record may not be gaps at all. "Gaps" presuppose the existence of "fill." In this case, the fill would be intermediate forms. But none exist; hence, neither do gaps.

Perhaps the best documented case of the punctuational mode of evolution is the Peter G. Williamson 1981 study of fossilized snails in the Lake Turkana region of Africa. The thick, undisturbed fossilized beds contain millions of preserved shells representing at least 19 species of snails. Several lineages of Cenozoic snails were exceptionally stable for 3 to 5 million years. When morphological changes in shell shape did occur, they were concentrated in brief periods of 5,000 to 50,000 years. These rapidly evolved new populations then persisted virtually unmodified until they became extinct. Some authors have cogently commented that so-called sudden changes within 50,000 years might appear to be instantaneous to a paleontologist but would be almost an infinity to an experimental geneticist. Indeed, to many geneticists, an interval of 50,000 years (or 5,000 generations in snails) would be sufficient time for the morphological changes in the snail populations to accrue gradually rather than dramatically.

"Hopeful Monster"

As previously discussed, there are genetic changes, known as *homeotic* mutations, which are responsible for such dramatic developmental effects as the transformation of one major organ into another. In the fruit fly, for example, an antenna can be transformed into a leg. Although this is fascinating for developmental studies, most evolutionists would not even contemplate that the origin of a new species could ever involve such drastic major morphological changes produced by homeotic gene changes, or *macromutations*. However, in the 1940s, the geneticist Richard Goldschmidt did propose that a new species of animal could evolve almost instantaneously by macromutations. He conceded that the vast majority of macromutations would have disastrous consequences ("monsters"). An occasional macromutation, however, might conceivably immediately adapt the organism to a new way of life, a "hopeful monster" in Goldschmidt's terminology.

Goldschmidt was convinced that the gradual accumulation of small mutations was too slow a process to account for broad macroevolutionary trends. He favored the idea of a dramatic, abrupt transformation of a population by a favorable, instantaneous macromutation, which could take the form of a radical rearrangement of genetic material (a gross chromosomal change). The concept of punctuated equilibria, as formulated by Gould and Eldredge, does not require or imply fortuitous favorable macromutations. Rather, in the process of allopatric speciation in peripheral isolates, the incipient species develop their distinctive features rapidly.

Evolution of Horses

In reconstructing the phylogeny of an animal group, it has been the standard practice to show the emergence from a central, generalized stock of a large number of divergent branches or lineages. Not all branches persist; indeed, the general rule is that all but a few perish. The disappearance of many branches in the distant past might lead the observer today to the mistaken impression that the evolution of a particular group was not at all intricately forked. Thus, the evolution of horses could be erroneously depicted as an undeviating, straight-line progression from the small, terrier-sized *Hyracotherium* (formerly known as *Eohippus,* the "dawn horse") to the large modern horse, *Equus.* On the contrary, a wealth of evidence has convincingly revealed a pattern of many divergent lineages (fig 19.3).

How does one interpret the divergent lineages? When new fossil discoveries revealed that a single, continuous lineage was implausible, proponents of phyletic gradualism simply postulated the existence of more than one line of gradual descent. We may dwell on one important lineage. The appearance of the one-toed condition was a landmark in horse history. The fleet-footed, one-toed *Pliohippus* emerged during the Pliocene epoch from the slow-footed, three-toed *Merychippus* (fig. 19.3). This major event is usually interpreted as a gradual change within a single lineage. In other words, a direct ancestral-descendant relationship between *Merychippus* and *Pliohippus* has been widely accepted. The paleontologist George Gaylord Simpson departed from the view of slow and steady change when he suggested that such a major transition might have involved the accumulation of small genetic changes in relatively

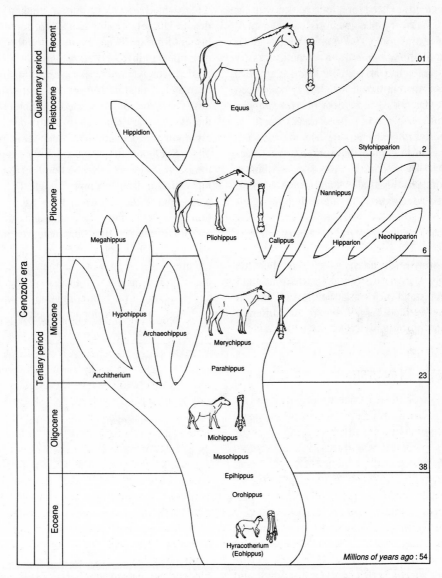

Figure 19.3 Phylogenetic tree of horses through time, with its many divergent branches. All branches died out, save one which eventually culminated in the modern group of horses, *Equus*. The history of horses dates back to early Cenozoic times, some 60 million years ago. (The Cenozoic era is divided into periods, which in turn are subdivided into epochs.)

rapid succession. This accelerated pace of phyletic gradualism was called "quantum evolution" by Simpson.

A geologically sudden appearance of *Pliohippus,* as prescribed by the model of punctuated equilibria, is also plausible. One might argue that *Merychippus* existed for a long period with little or no evolutionary change (stasis) and eventually suffered extinction. *Pliohippus* is not a linear descendant of *Merychippus* but a dramatically new form that earlier had arisen as a small peripheral population of the parent species. It may be noticed (fig. 19.3) that not all offshoots of *Merychippus* evolved the progressive single-toed condition. Several lines of Pliocene horses, such as *Hipparion* and *Nannippus,* retained the conservative three-toed pattern. These conservative Pliocene horses were evolutionary blind alleys and may well have represented "punctuations" that failed.

MODES IN HUMAN EVOLUTION

In the past two decades, Africa has become increasingly the focus for investigations into human origins and differentiation. Several sites in Africa have yielded fossils that reveal different phases of the human story from the earliest stages. The differences of opinion that exist concerning the phylogeny of the Hominidae are vividly revealed by the ever-changing reconstructions of the human and evolutionary tree, almost on a monthly basis (see chapter 17). One of the most widely held views is that a gradual ancestral-descendant relationship existed for each successive hominid group. Thus, the australopithecines shaded imperceptibly into *Homo habilis,* who in turn graded slowly into *Homo erectus,* with the latter ultimately transforming into modern *Homo sapiens*. The theory of punctuated equilibria claims that the evolutionary changes in hominids occurred in rapid bursts, concentrated in speciation events (fig. 19.4). Such punctuated episodes were separated by long periods of stasis in the hominid lineages during which little or no morphological change took place. Figure 19.4 shows that this pattern of evolution appears "rectangular" rather than tree-like.

Evolution through punctuated equilibria is a reasonable but controversial hypothesis. It remains to be seen whether the fossil record for many organisms does, in fact, show that the punctuational mode plays a comparatively major role in the origin of evolutionary novelty.

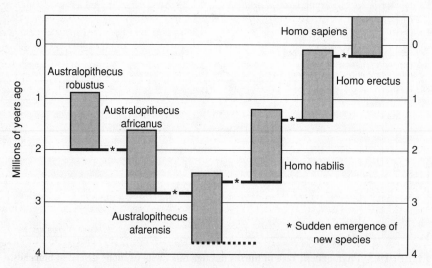

Figure 19.4 The rectangular pattern of hominid phylogeny as expressed by the controvertible theory of punctuated equilibria.

20

NATURAL SELECTION AND SOCIAL BEHAVIOR

In his provocative book entitled *Sociobiology*, Edward O. Wilson of Harvard University delves imaginatively into the social interactions of all animals, including humans. The cardinal theme is that the social interplay of animals, no matter how complex, has evolved by natural selection. Accordingly, sociobiologists attempt to place social behavior on sound Darwinian principles. The doctrine of natural selection is thought of as central to the biological understanding of sociality.

If the behavioral attributes of animals are the products of the same evolutionary forces that shape morphological and physiological traits, then social behavioral patterns should be adaptive. That is to say, behavioral dispositions should be optimally designed to confer reproductive success upon the individual. There are, however, behavioral acts that are detrimental to the individual performing the act but promote the reproductive advantage of other members of the population. A female worker bee, for example, completely forsakes procreation and labors ceaselessly to enhance the reproductive fitness of the queen. Is it possible to select for behaviors that are individually disadvantageous but beneficial to the species as a whole? This is one of the searching questions in sociobiology today.

GROUP SELECTION

In 1932, the English geneticist J. B. S. Haldane speculated on the possibility that a trait may be selected that confers an advantage for the group but is costly to the individual. He used the term *altruism* for such a trait and defined an altruistic act as one that decreases the personal fitness of the organism performing the act but is beneficial to the population as a whole. Thus, a prairie dog that emits a loud warning call (or alarm) when it spots a coyote improves the chances that his fellow conspecifics will survive. In protecting the group, the prairie dog (alarmer) attracts attention to itself and places itself in immediate danger of being captured by the predator. If the prairie dog were to remain silent, its presence to the coyote would not be betrayed. The warning act by the alarmer is essentially self-sacrificial; the term *altruism* may be equated with self-sacrificial or selfless. Haldane acknowledged the difficulty in explaining, in a mathematical model, the establishment of an altruistic trait that apparently diminishes one's own chances of survival and reproduction.

In 1962, the Scottish ecologist V. C. Wynne-Edwards evoked the concept of *group selection* to account for altruism as it relates to territorial behavior. Wynne-Edwards views territoriality as a method of population control. When a given population becomes excessive, many individuals cannot find territories and therefore cannot breed. Wynne-Edwards suggests that the territorial system has evolved by natural selection as part of a mechanism to stabilize the population density at a level that can be supported by the available food resources of the area. Wynne-Edwards' thesis assumes that natural selection operates for the benefit of the group as a whole. The implication is that many individuals are genetically or internally programmed to refrain from reproducing so as to not endanger the welfare or survival of the stock. In other words, natural selection has fostered both territorial "winners" and "losers" for the good of the species as a whole. Stated another way, a "loser" promotes the reproductive advantage of the "winner" at its own expense.

The curtailment of reproductive activity in "losers" is inconsistent with the notion of individual selection. An abridgment of reproductive behavior certainly does not benefit the individual. Natural selection characteristically operates on individuals (*not* groups) to augment (*not* decrease) individual reproductive capability. Since selection presumably operates solely to maximize the reproductive success of each individual, it is difficult to imagine how individuals can be selected to save the group at their own individual expense.

From a theoretical standpoint, can reproductive curtailment or restraint on the part of an individual evolve in a population so as to confer a reproductive advantage upon the group as a whole? Let us suppose that gene a_1 promotes reproductive capability and its allele, a_2, tends to curtail the reproductive capacity of an individual. Can the a_2 allele persist, or even spread, in a population when its effect is to impair the reproductive fitness of its possessor in the present and succeeding genera-

tions? The a_2 allele necessarily is selected against, since the possessors of the alternative allele, a_1, obviously leave more offspring than the possessors of the a_2 gene. Ultimately, the self-sacrificial a_2 allele will be replaced by the reproductively advantageous a_1 gene.

KIN SELECTION

Altruism apparently cannot evolve without violating the principle of individual selection, since the individuals displaying self-sacrificial behavior are less biologically fit than their selfish (or egoistical) colleagues. Yet altruistic behaviors are clearly evident, particularly among birds and mammals. A female bird behaves altruistically when she protects her brood against predation; a male baboon will emerge from the heart of a troop to attack a leopard that threatens the group; and a human will place his or her life in jeopardy to rescue a drowning person. In 1964, W. D. Hamilton of the London School of Economics proposed an alternative route for the evolution of altruism that is not founded on the indefensible premise of group selection.

Hamilton explained altruism as the outcome of a selective process called *kin selection*. The concept is based on the fact that close relatives share a high proportion of the same genes. Altruistic behavior can be favorably selected if the probability is high that the beneficiaries of the altruistic act also have the same genes as the self-sacrificial altruist. Under this view, kin selection is a special manifestation of *gene* selection. It is the gene coding for a particular behavior that is optimized or favorably selected. A given allele a_1 will spread if the behavior associated with this gene adds a greater number of a_1 alleles to the next generation than in the preceding generation. This can happen only if the reproductively successful individuals are close relatives of the altruist, thereby increasing the probability that they carry the same a_1 allele.

An instructive phenomenon is the special chirp, or warning call, in bird populations, by which one member alerts the flock to the danger of a predator.

We may assume that the warning note is governed by an a_1 allele. A warning call originating as a signal from a parent to its offspring during the breeding season is selectively advantageous, because the degree of relatedness between the caller and recipient is high. Gene a_1 will persist in the population, since the alarm call by the parent (bearing gene a_1) increases the fitness of sufficient numbers of offspring also possessing a_1. By protecting the offspring, the parent invests in its own genetic representation in subsequent generations. In an evolutionary sense, the offspring that are protected by the parent are, in part, genetically the parent itself.

The lifetime of an organism is to be viewed as a strategy for perpetuating the organism's genes. An individual can transmit its genes directly through its own reproduction as well as by proxy through its close relatives who share genes by common descent. An individual is said to maximize an *inclusive fitness*—its own reproductive fitness plus the reproductive fitness of close relatives. The degree of genetic relatedness has a crucial bearing on the likelihood that two individuals will behave altruistically toward one another. The extent to which two individuals share genes by descent from a common ancestor is referred to as the degree of relatedness, or coefficient of relationship (designated r). In diploid species, the degree of relatedness of an individual to his or her full siblings is 1/2; to half-siblings, 1/4; to children, 1/2; and to first cousins, 1/8. Stated another way, a given individual shares 50 percent of his genes with a full brother or sister but only 12.5 percent of his genes with a first cousin. By this logic, grandparents and grandchildren are related by 1/4.

Kin selection molds a form of altruism that is channeled to genetic relatives. An altruistic gene will be perpetuated if the reproductive benefit gained by the recipient of the altruistic act exceeds the cost in reproductive fitness suffered by the altruist. Figure 20.1 depicts kin selection with respect to two brothers. The altruist leaves no offspring, but his sacrificial act enables his brother to leave more offspring than he would have otherwise. The altruist's genes have been effectively removed from the population by his failure to reproduce, but the genes

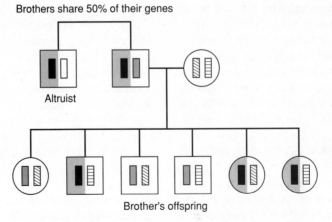

Brothers share 50% of their genes

Altruist

Brother's offspring

Figure 20.1 Sibling altruism. The coefficient of relationships between two brothers is 1/2. Thus, the probability is 0.50 that a gene (or chromosome) present in one individual also occurs in his brother. For simplicity, only a pair of chromosomes is depicted. The individual's altruism to his brother is selectively advantageous because the altruistic act enables his brother to transmit greater numbers of replicas of the shared chromosome to the next generation than the individual himself might have transmitted.

that he shares with his brother are more than restored by three of the offspring sired by his brother. The altruist has actually gained genetic representation in the next generation. Indeed, he has gained in *inclusive fitness,* even though he has lost fitness in the classical sense.

Kin selection provides an explanation for the baffling social behavior in ants, bees, and wasps. Curiously, whole castes of sterile females devote their entire existence to the welfare of the queen. Females are diploid individuals that develop from fertilized eggs with maternal and paternal sets of chromosomes. Males are the haploid products of unfertilized eggs and possess only the maternal set of chromosomes (fig. 20.2). The startling outcome is that the sibling daughters of a queen are more closely related to each other than they would be to any of their own daughters!

Specifically, female workers share three-fourths of their genes with their sisters ($r=3/4$). This is the case because sisters share the same set of paternal genes and share additionally, on the average, one-half of the maternal genes (fig. 20.2). Thus, sisters are related by the average of 1 (for paternal genes) and 1/2 (for maternal genes), or 3/4. If a daughter were to produce her own offspring, that daughter would share only one-half of its genes with any of her offspring. Accordingly, a female worker actually contributes more to her Darwinian fitness by assisting her mother in raising offspring (3/4 relationship) than by rearing her own offspring (1/2 relationship). Natural selection has fostered an unusual form of cooperative behavior among sisters.

The male bees, or drones, show a pronounced lack of concern over the welfare of their sisters. This is not surprising, since drones share only 1/4 of their genes with their sisters. Since a drone possesses only maternal genes, a sister cannot share any paternal genes with her brother (fig. 20.2). The total genetic relationship of a sister to her brother is the average of zero (for paternal genes) and 1/2 (for maternal genes), or 1/4. The drone is notoriously lazy ("lazy as a drone"), since his inclination is to produce daughters (who share all of his

genes) rather than help his sisters. In turn, sisters invest more energy in raising sisters than brothers and characteristically drive their brothers from the hive in the early summer. The selfishness of the drones and the diligence of the female workers are predictable aspects of kin selection.

KIN SELECTION IN HUMANS

A variety of altruistic behavioral dispositions in the human species—the sharing of food, the sharing of implements, and the caring for the sick—probably evolved by kin selection in the early hominid hunting bands. These bands consisted largely of tightly knit groups of close relatives. In fact, in primitive cohesive societies, social behavior appears to have been dominated by kin selection.

A striking phenomenon in some human societies is the strong parental attitude exhibited by a brother to his sister's children, far out of proportion to his degree of relatedness to his nephews and nieces. However, as predicted by kin selection, a brother's sense of responsibility to his sister's children increases as confidence in paternity of his own children diminishes. A brother is always genetically related to his sister's children, whereas he may be totally unrelated to his wife's children if the probability of paternity is low. Apparently, the certainty of a genetic relationship between brother and nephew (or niece) outweighs the dubious or tenuous relationship between father and alleged son (or daughter). In essence, a high probability of paternity is a necessary antecedent for extended parental investment by the male.

It has been suggested that homosexual behavior is a product of kin selection. Although the genetic factors for homosexuality lower reproductive fitness, the production of offspring occurs by proxy through the homosexual's immediate kin. In this vein, homosexuals may be compared to sterile worker bees who help raise their close genetic relatives. The behavior of the homosexual increases the fitness of genetic relatives more

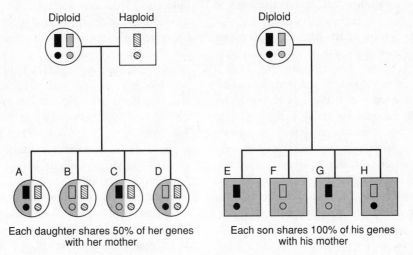

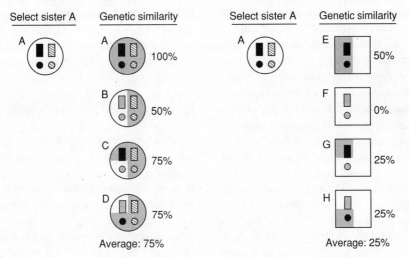

Figure 20.2 Degree of genetic relationship of a female bee to her ofspring *(top)* and the genetic relatedness of sisters to sisters *(lower left)* and sisters to brothers *(lower right)*. Females develop from fertilized eggs and have two sets of chromosomes (diploid state), whereas males develop from unfertilized eggs and have only one set of chromosomes (haploid state). For simplicity, the haploid set is represented by only two chromosomes. Sisters are more closely related to one another than they would be to their own offspring. The coefficient of relationship between mother and daughter is 1/2, but the coefficient of relationship between two sisters averages 3/4. Sisters have relatively few genes in common with their brothers, the coefficient of relationship averaging 1/4.

than it decreases his or her own fitness. This view is consistent with kin selection, but there are as yet no supporting or confirmatory data.

Kin selection may have shaped the phenomenon of menopause in the female. The middle-aged woman enhances her fitness by caring for her children's children. As the menopausal woman loses her capacity to have her own children, she regains inclusive fitness by devoting her efforts to the rearing of her grandchildren. It is interesting that the duration of fecundity is greater in women of economically advanced countries. Menarche occurs earlier and menopause is deferred. With improved nutrition and better personal hygiene, the female increases the duration in which she can invest in her own children.

THE SELFISH GENE

The alarm calls of birds are often cited as an excellent example of kin selection. It is generally the case that a close kin is sufficiently near the caller to benefit from the warning call. However, some authors contend that the habit of warning cries evolved by conventional individual selection rather than by kin selection. A warning call may *not* serve to draw the attention of the predator to the alarmer. The warning call may *not* expose the caller to greater danger. Rather, the call note is intended to mislead the predator as to the position of the prey.

Under this view, the alarmer utters a call that the predator cannot easily locate. Since the predator cannot readily detect the location of the call, it would appear that the alarmer issues the warning note to protect itself and not its colleagues (conspecifics). Suppose a flock of birds is feeding in a meadow and a hawk flies past in the distance. The hawk has not seen the flock, but there is the danger that he will soon spot the flock. The noisy rummaging activity of members of the flock could attract the hawk's attention. The alarmer who first detects the hawk issues a warning call to its companions but does so only to curtail the noise of the flock. Accordingly, the caller reduces the chance that the flock will inadvertently summon the hawk into his own vicinity. In large measure, the caller is not acting *altruistically* but rather *selfishly*.

The same situation may be viewed in another selfish vein. Suppose the caller fails to signal, quietly taking flight by itself without warning its unsuspecting conspecifics. In this event, the hawk might be easily attracted to a single bird flying off. The issuance of an alarm would be better than no alarm at all. An alarmer that issues a call curtails his risks by flying up into the tree—and simultaneously arouses the other members to fly with him into the tree. In corralling the conspecifics, the caller ensures protective cover for himself in flight. The conspecifics benefit from the caller's act, but in increasing his own safety, the caller benefits even more!

The portrait of a biological individual that emerges from these examples is one of self-serving opportunism. It seems that an individual is programmed to care about itself. It may be, as several investigators contend, that acts of apparent altruism are actually selfishness in disguise.

RECIPROCAL ALTRUISM

When a self-sacrificial parent saves his own child from some potentially tragic event, the parent acts to protect his share of genes invested in the child. If, however, an individual saves the life of an unrelated party or nonrelative, kin selection is ruled out. In this instance, the individual may have behaved altruistically with the expectation that the beneficent behavior will be reciprocated by the stranger at some future occasion. A person who saves another person from drowning, for example, hopes in turn for help when he or she is in danger. This is the concept of *reciprocal altruism*, first proposed in 1971 by Robert L. Trivers of Harvard University. An altruist who places himself in danger for a biologically unrelated person incurs a reduction in fitness because of the energy consumed and the risks involved. However, a future reciprocal act by the recipient is likely to bring returns that are equal to, or greater than, the altruist's original expenditure.

There is definite survival value to both donor and recipient when altruistic acts are mutually exchanged. Each participant benefits by increasing his or her fitness. If, however, a recipient fails to reciprocate when the situation arises, that recipient would no longer have the benefits of future altruistic gestures to him. The nonreciprocator would then be at a selective disadvantage since subsequent adverse effects on his life would not be overcome by altruistic acts that might be lifesaving. The ultimate effect of nonreciprocation would be the restriction of altruism to faithful fellow altruists, with the consequence that genes for reciprocal altruism would be perpetuated through the generations. In fact, the chances of selecting for altruistic genes are improved as more altruistic acts occur in the lifetime of the altruist. By the continual exchange of beneficent acts, altruists accrue more fitness in the long run than the nonreciprocators. Overall, altruists gain fitness, rather than lose fitness, from selfless acts.

"Stepmothering" may be considered a form of reciprocal altruism. When the biological mother dies or leaves her children, the male may remarry, and the stepmother assumes the care of the offspring. The basis of the reciprocity is best seen in the behavior of western gulls off the coast of California. An unmated gull, by raising the offspring of a previous mate, establishes a pair-bond relationship with the male and guarantees herself both a mate and a breeding site for subsequent years. Thus, in a population of gulls where there is a surplus of breeding-age females, the acquisition of a breeding territory complete with a mate is a compelling incentive for a female gull to behave altruistically by rearing another female's offspring on one occasion. By so doing, she increases her personal reproductive fitness for future years with only a relatively small initial investment. Such behavior is fostered by natural selection, since the final effect for the stepmother is an increase in her own genetic material in succeeding generations.

PARENTAL REPRODUCTIVE STRATEGIES

In the placental mammals, the young develop slowly and the prolonged period of caring for the young is almost exclusively restricted to the female. The males of most mammalian species do not establish even a semblance of a permanent relationship with the pregnant and nursing female. In its extreme form, such as in cats and bears, the female actually ejects the male from any contact with the young. It is as if the mammalian male were biologically superfluous after insemination. Apart from insemination, most mammalian females have little use for the males—except possibly to protect them from other males. It is only exceptionally, as in wolves and humans, that the male plays an important role in the protection and care of the mother and child.

Each sex, without conscious intent, strives to attain maximal reproductive success. The reproductive strategies of males and females are clearly different. Each sex may be viewed as an investor in which the capital invested is in terms of reproductive effort. The female undoubtedly invests more per offspring than the male. The overproduced, lightweight sperm are inexpensive compared with the thriftily produced, energy-rich eggs. The prolonged internal gestation and the protracted period of maternal care of the newborn place rigorous demands of energy and time on the female. Accordingly, the female has a much greater stake in any one reproductive act. Since a reproductive mistake is much more severe for the female, she is more discriminating than the male in the choice of a mate.

A male's basic reproductive strategy is to achieve as many fertilizations as possible. Males are less selective in the choice of acceptable sexual partners and more aggressive in excluding other males from the opportunity to mate. Females are choosier and carefully evaluate their options. Males almost universally act competitively as "sexual advertisers," and females act cautiously as "comparison shoppers." The task of critical discrimination of a sexual partner and the avoidance of an unproductive pregnancy resides almost wholly with the female.

The female selects that male as a sexual partner whose appearance and behavior signify that he will transmit a superior set of genes to the offspring. Mammals are generally polygynous—that is, a single male mates with more than one female. As previously mentioned, most mammalian males do not establish any durable association with the female beyond the sexual act. If a reproductively fit mammalian male has little to offer except his superior genes, then it is to the female's advantage to mate with that suitable male, no matter how often he may have already mated. In polygynous mammalian species, the fittest males (at least those that are behaviorally the fittest) have the greatest reproductive success. For example, in breeding colonies of elephant seals, the males clash for dominance status, which is associated with mating rights. Less than one-third of the males in residence during a season copulate, and as few as five dominant males account for 50 percent of the mating in a given season. Clearly, females who share an oft-mated male of unusual competitive ability are attracted to a "winner" with "proven" genetic qualities.

The situation is different with birds, in which monogamy is the general rule. One male and one female form a breeding pair and remain together, at least for the rearing of one brood. In some species of birds, the same pair-bonding may persist for several successive broods (as in songbirds) or even for a lifetime (as in geese and swans). In these situations, the male's contributions to the female and her offspring are much more than a mere complement of genes. The males of many species of birds defend a territory, provide protection against predators, assist in the building of nests, incubate the eggs, and furnish nourishment for the female and her young. Thus, where the male provides numerous indispensable services, it is to the female's advantage to maintain pair-bonding. Stated another way, we can expect enduring pair-bonding only where the male plays a substantial paternal role.

The tendency toward polygyny in human males apparently has been dampened by strong selective pressures. The human male engages in an elaborate courtship ritual, which is in itself a substantial investment for sustained pair-bonding. Robert Trivers has suggested that the prolonged male courtship serves as insurance for the male that he is not establishing a bond with an already inseminated female. The greater the male investment, the greater the importance of paternity knowledge. To assure accurate identification of his own offspring, the male jealously guards against "cuckoldry"—that is, against the possibility that he might invest in offspring that are not his own. Human males universally treat their mates as possessions and sequester them from the adulterous interests of other males. It is only when the female remains faithful that the male can overcome his ever-present uncertainty about paternity. Nevertheless, the other side of the male's strategy is his inclination to cuckold other males. There have evolved, however, female counterstrategies to improve the female's ability to hold her mate's attention and care.

With the notable exception of humans, the female of mammalian species becomes sexually aroused only at certain seasons. At specific times the female comes into heat, or *estrus,* and only during these restricted periods is the female receptive to the male. The onset of heat is typically synchronized with ovulation, or the release of the ripe egg. In simultaneously advertising sexual receptivity and ovulation, the mammalian female provokes intense competition among the dominant males and increases her likelihood of securing a male of exceptional genetic constitution. The females are monopolized by the males only at ovulation.

In the human female, there are generally no outward or conspicuous signs of ovulation. Human females may be sexually receptive during any part of the ovulatory cycle. Several authors have pointed out that the loss of the phenomenon of estrus in human evolution was an event of great importance. The concealment of ovulation by the female serves to maximize her mate's confidence of paternity, in that potentially competing males are not flauntingly apprised of the ovulation event. Stated another way, if the human female were to glaringly advertise ovulation, it might cost her the protection and parental care of her mate by attracting competing males who threaten her mate's confidence of paternity. Moreover, her potentially philandering mate is restrained in seeking copulation with other fertile females who, like her, conceal the event of ovulation.

When the pair-bonding relationship fails or is dissolved, the female typically experiences difficulty in finding a substitute male to aid in the rearing of the offspring. We would expect strong selection pressure against "stepfathering," since the substitute father is called upon to invest in some other male's genes. The literature is replete with examples in varied mammals in which a male eliminates the offspring of a previous mating and replaces them with his own offspring. When the male lions of a new group depose the resident males of a pride, the new regime often kills the cubs of the displaced males. In another instance, a pregnant female mouse responds to the odor of a novel male by aborting her fetus and returning to estrus in preparation for propagating the genes of the new male. Among the Yanomama Indians of Venezuela, a man will order his newly acquired wife to kill her infants of a previous marriage. The biological urge to invest only in one's own offspring is evidently powerful and pervasive.

PARENT-OFFSPRING CONFLICT

Robert Trivers defines parental investment as the amount of care and assistance rendered by the parent that increases an offspring's chance of future reproductive success at the cost of limiting the parent's capacity to invest in other offspring. Each offspring is of equal value to a parent, since 50 percent of the parent's genes are distributed in each offspring. But each offspring is completely related to itself ($r = 1$), and related by one-half to each of its sibling ($r = 1/2$). Each offspring, then, would expect to receive twice as much parental care for itself as for each of its siblings. Thus, it is inevitable that conflicts will arise between parent and offspring.

In particular, a mother will normally apportion her investment among her offspring to maximize the number that survive to reproductive maturity.

She must necessarily limit the amount of care that she can give to any one child. On the other hand, if the child is to maximize its own chances, the child will attempt to extract more than its share of parental care. The parental investment in an offspring eventually reaches a point of diminishing returns. That point is reached when the mother's investment in a given child costs her more than she gains and where it costs the child more than the child gains. Once the point of diminishing returns is met, the parent-offspring bond should be severed.

IMPLICATIONS FOR SOCIETY

Some writers declare that any behavioral act that increases biological fitness (individual or inclusive) is selfish and exploitive. The psychologist Donald Campbell of Northwestern University decries that man's basic behavioral tendencies are self-centered. Campbell advocates that society must counter the individual selfish tendencies that are promulgated by natural selection and, through learning and socialization, promote positive or beneficent expressions of social interplay. The implication is that our genetically based social tendencies are "bad" and only culturally imposed restraints are "good." Campbell presupposes that a social system works best by placing restraints on individual biological fitness. He also presumes that biological evolution and cultural evolution are sharply separated from one another. He thus revives the old, untenable nature-nurture dichotomy. This dichotomy is a trap of extremes. The extreme pessimistic view is that human behavior is solely influenced by genes, and any attempts to change behavior by learning are futile. The other extreme is the misguided optimism that humans are so infinitely plastic in social behavior that they can be molded easily to ideal social and economic systems.

In reality, the complex social behavior pattern of the human species is the outcome of the complementary interaction of biological evolution and cultural evolution. The key question is not whether there is a genetic foundation of social behavior in humans, but rather how firm and constraining is the foundation. The greater our knowledge and understanding of the degree of genetic determinism, the greater our capacity to modify behavior by cultural changes.

Richard D. Alexander of the University of Michigan asserts that we need not summarily suppress our biological inclination. Human beings *do* behave in such a way as to maximize their genetic representation in future generations. Indeed, Alexander suggests that human culture is largely the manifestation of the reproductive strivings of the individual members of societies. Individuals, in maximizing their inclusive fitness, favor more closely related kin over more distantly related individuals. An individual's inclusive fitness is not generously served by aiding an unrelated person or stranger. An individual might aid a stranger on the expectation that such a person would ultimately, at some future date, reciprocate the aid in a similar circumstance.

The foregoing pattern of behavior is exemplified by a familiar example in our daily lives. We *felicitously give* Christmas presents to our close relatives. We *occasionally make* a gift to our more distant relatives. We *charitably exchange* Christmas cards with our acquaintances. Finally, we *unsparingly remove* from our list of card exchanges those acquaintances who fail to reciprocate.

CULTURAL EVOLUTION

Some 10,000 years ago, humans gave up the precarious hunting and gathering way of life for a more settled and secure existence based on agriculture and the breeding of animals. This initial trial at cultivating the land ushered in the so-called Neolithic revolution. When we domesticated plants and animals, we took a major step in controlling nature rather than being at its mercy. We placed nature at our service with our ideas, discoveries, and inventions. We began to control our own food supplies, to congregate into more stable communities, and to establish a distinctive civilization.

We have since modified our external surroundings at an ever increasing rate. We have undergone an industrial revolution and are now witnessing a technological revolution. The rapidity and efficiency with which we have dominated the environment reflects our capacity for learning and transmitting our accumulated knowledge. This capacity for *cultural evolution* has had a profound influence on our way of life and on our destiny.

CHARACTERISTICS OF CULTURAL EVOLUTION

There are important differences between biological and cultural evolution, one of the primary differences being the tempo of change (table 21.1). The brisk pace of cultural evolution contrasts sharply with the gradual progress of biological evolution. The acquisition and transmission of learned ideas occurs through the generations from mind to mind rather than through the germ cells. Cultural transmission is rapid, limited only by the efficiency of communication methods and our inventiveness.

A given cultural change can be passed on to large groups of unrelated individuals, whereas a given genetic modification can be transmitted only to direct descendants, who are generally few. Biological evolution is necessarily slow since it depends on accidental mutational changes in the DNA molecule, and each chance genetic change may take many generations before it can become established in the population under the force of natural selection.

TABLE 21.1	Comparisons of Biologic and Cultural Evolution

	Biologic Evolution	Cultural Evolution
Agents	Genes	Ideas
Rate of change	Slow	Rapid
Direction of change	Random mutations; subject to selection	Usually purposeful and directional
Nature of new variants	Often harmful	Often beneficial
Transmission	Parents to offspring	Wide dissemination by many means
Distribution in nature	All forms of life	Unique to humans

PROCESS OF CULTURAL EVOLUTION

Cultural evolution does not supersede biological evolution; rather, it interacts with biological evolution. As a striking example, population geneticists have noted that the incidences of two principal congenital malformations, *spina bifida* (literally, "two-part spine") and *anencephaly* (virtual "absence of brain"), have declined in recent years in certain areas of the world, notably in England and Wales. These two abnormalities are but part of a group of malformations associated with the embryonic neural tube, the future brain and spinal cord. The cluster of abnormalities is presently referred to collectively as *neural tube defects.* One of the more serious defects is a *myelomeningocele,* a cyst-like sac that contains a displaced portion of the spinal cord as well as disorganized spinal nerves (figure 21.1).

Genetic factors are instrumental in causing neural tube defects; this is indicated by the fact that there are twice as many afflicted offspring in marriages of first cousins as in marriages of unrelated persons. Moreover, both disorders exhibit a familial concentration, in that more than one sibling is often affected. The recent decline in neural tube defects in the British Isles and other modernized countries has been attributed to several important social and medical considerations. The first is the increased awareness of the familial nature of the disorders, which has caused many

parents of an affected child to avoid subsequent pregnancies. The second is the increased acceptance and effectiveness of contraceptive measures, which has enabled parents to restrict family size more successfully. The third is the medical availability of prenatal screening tests, by which the neural tube defect can be detected *in utero.* In terms of biological evolution, the long-term impact of deliberate family limitation practiced on a large scale would be a reduction in the pool of detrimental genes responsible for these neural tube disorders. Evidently, social or cultural considerations, or changes in one's way of life, can influence the direction of biological evolution.

The foregoing measures should not be construed as indicating a retrenchment by medical researchers from the search for *preventive* innovations. As we have learned, genes do not act in a vacuum. Environmental factors also play a role in the etiology of neural tube defects. For more than two decades, the claim had been made that folic acid, an important chemical in the biosynthesis of DNA, would reduce the rate of occurrence of neural tube defects. Today, it is generally conceded that the dietary intake of folic acid by women may safeguard their fetuses against neural tube defects. The protective effect has been seen with doses ranging from 0.4 milligrams per day to 4 milligrams per day. The recommendation is that all women take folic acid prior to conception and throughout pregnancy.

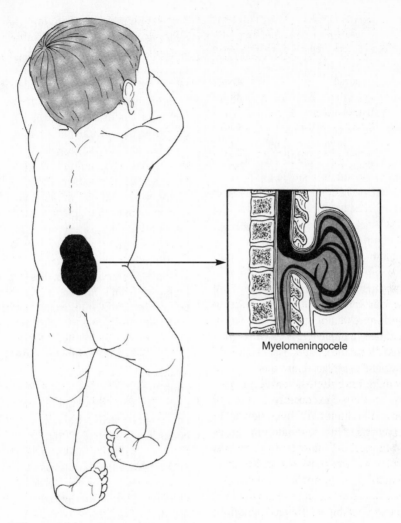

Figure 21.1 A myelomeningocele is essentially a herniation of the spinal cord. The cyst-like sac contains spinal nerves that cannot be repaired.

Preconceptual use is important, in that the formation of the embryonic brain and spinal cord occurs during the first few weeks postconception—a time when many women are unaware that they are pregnant. It is estimated that more than half of the worldwide incidence of neural tube defects could be prevented if all women took folic acid supplementation.

Each new medical thrust presents acute challenges to existing moral codes and conventional wisdom. In anencephalic fetuses, the cerebral hemispheres are completely missing or reduced to small masses attached to the base of the skull. Anencephalic infants invariably are stillborn or die shortly after birth. Notwithstanding a deteriorated nervous system, the remaining organs of the body are remarkably well-formed. The question being asked, given the serious shortage of donor organs, is whether it is ethical to use the organs of anencephalic fetuses for transplantation.

GENE THERAPY

There is no effective treatment for most severe genetic disorders. Some, like Tay-Sachs disease, follow a relentless course to early death; others, like Lesch-Nyhan syndrome, are crippling painful disorders that linger for many years. Debilitating and often fatal inherited disorders are caused largely by coding errors in the gene. In many cases, only one miscoded nucleotide can trigger a major disorder. As a result of a single faulty gene, the body may fail to manufacture an important protein or enzyme. The absence or deficiency of a vital chemical substance may lead to the malfunctioning of several different bodily processes.

When an infant is born with a serious inherited condition, the physician can do very little in terms of therapy. Existing treatments are more palliative than curative. However, in years to come, some life-threatening genetic disorders may be corrected by a procedure known as *gene therapy*. In theory, the procedure is deceptively simple; a faulty gene is replaced, or compensated, by a properly functioning gene. In reality, the task is formidable, and the gene, once incorporated, must be able to function properly in its new location without causing harm. Despite numerous impediments, the current pace of research is rapid, and the treatment of human disorders clinically by gene therapy is within the grasp of medical scientists.

"Gene therapy" is a highly charged, emotional expression. The apprehension in the general public is that human genes will be manipulated to remake or change the nature of human beings, like Dr. Frankenstein's monster. The public fears the exercise of inordinate power by scientists. Such anxiety is misplaced, reflecting in large measure the public's lack of understanding of the motivations and goals of scientists. The objective of biomedical research is, and has always been, to alleviate human suffering. The focus of gene therapy is on debilitating, relatively rare inherited disorders, primarily enzyme deficiencies caused by simple mutations. Even when the techniques become improved, only a few of the 7,000 known single-gene disorders are likely to be treatable. Indeed, the requisite biochemical basis is scarcely understood for most genetic disorders. Since the actual repair of a faulty human gene *in situ* is still in experimental stages, the current strategy entails the insertion of a functional gene to offset the effects of the detrimental gene.

The initial step in gene therapy is to isolate the normal counterpart of the gene responsible for the disease and produce many copies of the normal gene. One way investigators can isolate a human gene is by taking advantage of the fact that genes are the repositories of information for proteins. As earlier discussed, the gene (DNA) does not directly form a protein but works through messenger RNA. A particular messenger RNA can be extracted from a human cell (globin *m*RNA from an erythrocyte, for example) and used as a template to synthesize a complementary strand of DNA, abbreviated *c*DNA. These single-stranded *c*DNA molecules are then converted into double-stranded DNA molecules, from which multiple copies, or *clones,* are made. Because introns are eliminated by RNA splicing, the *c*DNA contains an uninterrupted sequence of nucleotides that codes for the protein (fig. 21.2). However, the *c*DNA lacks the gene's regulatory regions—that is, the DNA sequences (promoters and enhancers) situated upstream from the gene's coding section without which transcription into messenger RNA is unlikely. Accordingly, as seen in figure 21.2, the regulatory regions are spliced onto the *c*DNA.

The therapeutic *c*DNA is introduced into its target cell by a delivery vehicle, or *vector,* which can be a bacterium or virus. Attention of late has centered on RNA viruses *(retroviruses)* as reliable vectors for transfer. A retrovirus is remodeled into a delivery vehicle by deleting much of the native genetic information of the retrovirus and replacing the viral sequences with the therapeutic gene sequences.

As shown in figure 21.2, the viral *env* gene is replaced by the cloned human gene and the viral *gag* and *pol* genes are replaced by a bacterial gene

Native virus

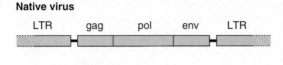

Remodeled virus

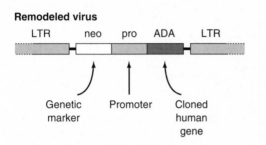

Figure 21.2 **A retrovirus remodeled to serve as a vector for gene therapy of adenosine deaminase deficiency.** The LTR is a "long terminal repeat" of sequences that include regulatory signals. In the native virus, *gag* is the gene for the core proteins of the virus, *pol* is the gene for reverse transcriptase, and *env* is the gene for the other envelope proteins of the virus. In the remodeled virus, *neo* represents the gene for resistance to neomycin (thus allowing the selection of cells that have been infected by the virus). *Pro* is the promoter sequence for the inserted human gene, and ADA is the normal, cloned human gene for adenosine deaminase.

(neo) that confers resistance to the antibiotic neomycin. The *neo* gene serves as a marker that allows investigators to readily identify target cells that have incorporated the modified retrovirus. Thus, when target cells are exposed to an analog of neomycin that is highly toxic to human cells, only those cells expressing the *neo* gene survive.

Additional modifications are necessary. The genome of the retrovirus has been so drastically altered that the remodeled version is unable to make the protein coat that is necessary to form a complete viral particle. To alleviate this problem, the remodeled virus is associated with a "helper" virus—a retrovirus that is streamlined to retain only the gene for the protein coat. The helper virus furnishes the protein coat for the remodeled virus,

enabling the latter to enter the target cell. Once inside the target host cell, the vector *c*DNA sequences become stably integrated into the host genome in the nucleus. The therapeutic gene is carried permanently in the host genome and is passed on to all progeny cells after cell division of the host cell. In its integrated state, the therapeutic gene is able to express itself, producing the desired protein product.

TARGETS FOR GENE INSERTION

One of the primary targets for gene insertion has been the hemopoietic stem cells of bone marrow, a self-sustaining population from which mature blood cell types are derived. One hemopoietic stem cell can generate up to one million mature blood cells. These pluripotent stem cells comprise less than 0.1 percent of the marrow cells.

The experimental strategy using animal models customarily begins by removing marrow tissue from the animal (fig. 21.3). The explanted marrow is infected in the laboratory with a packaged retrovirus carrying a copy of the gene of interest. The animal is then exposed to sufficient radiation to destroy all of its remaining bone marrow. In this system, the destruction of the animal's remaining bone marrow enhances the opportunity for newly introduced marrow cells to grow. After irradiation, the infected marrow cells are reintroduced into the animal, where they proliferate and repopulate the blood-forming system. The animals are tested to determine whether the new blood cells carry the transferred gene and whether the gene is expressing itself properly.

In 1984, Richard C. Mulligan and his associates at the Massachusetts Institute of Technology packaged a retrovirus to transfer the bacterial antibiotic-resistant gene *(neo)* to explanted bone marrow tissue of an irradiated mouse. If the stem cells were to incorporate the antibiotic-resistant gene introduced by the engineered viruses, the recipient mouse would possess the marker gene in every type of differentiated blood cell derived

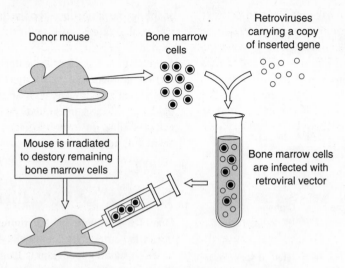

Figure 21.3 Gene transfer in a mouse model. Bone marrow
cells from a donor animal are infected in the laboratory with a retro-
viral vector carring copies of the exogenous gene. After infection,
the treated marrow cells are infused into a genetically indentical
recipient laboratory mouse whose hemopoietic system has been
destroyed by irradiation. In the absence of normal blood-forming
cells, the introduced infected marrow cells gain a competitive
advantage. If the experiment were to be performed in humans, the
treated cells would be injected back into the donor and the infected
marrow cells would have to compete with the population of marrow
cells that is normally present.

from the stem cells carrying the implanted gene.
In conformity with expectation, the antibiotic-
resistant gene was found to be present in differentiated white blood cells of the spleen. This finding
was clear-cut evidence of successful gene transfer
since the mature blood cells of the spleen of the
irradiated host mouse could only have originated
from the introduced hemopoietic stem cells. The
efficiency of gene transfer was only 20 percent—
that is, the recombinant viruses integrated only in
a subpopulation of the stem cells. Moreover, the
viruses could not be targeted to particular sites in
the chromosomes. Nonetheless, these results, as
well as others, are very encouraging.

The immediate hopes for somatic cell therapy
in humans revolve around those diseases that can
be treated by manipulating hemopoietic cells of

the bone marrow, since this tissue can be readily
removed, manipulated *in vitro,* and easily reintroduced into an intact host. One promising candidate
for gene therapy is *adenosine deaminase* (ADA)
deficiency, a rare disease of the immune system.
Victims are born without the gene for synthesizing
adenosine deaminase, an enzyme essential to the
production and maintenance of immunologically
important lymphocytes (T cells and B cells). With
the immune defense severely impaired, affected
children are predisposed to recurrent and persistent infection.

In ADA deficiency, even the production of a
small fraction of the normal enzyme activity (10
to 15 percent of normal) would be beneficial to
the patient. The hope is that enzyme production
can be fostered by placing the appropriate nor-

mal human gene via a virus into human patients. The protocol calls for removing the defective bone marrow from a patient, inserting a normal enzyme-producing gene into a number of marrow cells, and reimplanting the treated bone marrow into the patient. The same procedures are appropriate for several heritable immunodeficient disorders that are manifested primarily in bone marrow-derived cells.

THE REALITY OF GENE THERAPY

On the 14th of September 1990, banner headlines in the United States announced that a cheerful Ashanti DeSilva, barely four years old, was the recipient of the first federally authorized human gene transfer. She had inherited, from each parent, a defective version of the gene that normally synthesizes the aforementioned enzyme, adenosine deaminase (ADA). The heralded therapy was performed at the National Institutes of Health by a highly trained medical team, consisting of W. French Anderson, R. Michael Blaese, Kenneth W. Culver, and a host of others. White blood cells (lymphocytes) were removed from DeSilva's body and normal ADA genes were delivered into the pool of lymphocytes. The treated blood cells were then infused back into her bloodstream. This pioneering effort did produce therapeutic benefits. To this day, tests reveal that at least 50 percent of DeSilva's circulating lymphocytes, particularly T cells, contain a normal ADA gene.

Clinical trials of gene therapy have now been undertaken on a wide range of inherited disorders, including cystic fibrosis, hemophilia, familial hypercholesterolemia, and peripheral vascular disease. The most vexing problem is twofold—finding the best means of transferring normal genes to adequate numbers of target cells, and then having the confidence that the inserted gene will express its normal activity in the target cells. We shall consider below a novel procedure in which neither bacteria nor viruses are used as agents, or vectors, to transfer genes.

A DOSE OF NEW GENES FOR VASCULAR DISEASE

Surgeons have described a *popliteal artery aneurysm* as a sinister harbinger of sudden catastrophe. An aneurysm is a saclike dilatation of the wall of an artery, often at sites prone to atherosclerotic plaques. A common site of aneurysm formation is the popliteal artery, the major vessel feeding the arteries of the legs and foot (fig. 21.4). Popliteal aneurysms tend to be overlooked by both patient and physician and are recognized only when complications occur. The most serious complication is rupture of the thinning arterial sac, which requires immediate surgical intervention to save the limb and, occasionally, life. The customary operative treatment is bypass grafting, using the patient's own vessel (typically, the saphenous vein) as the bypass graft.

In 1998, Dr. Jeffrey Isner and his colleagues at St. Elizabeth's Medical Center in Boston University published an update on their ongoing research of a novel type of gene therapy in the treatment of peripheral vascular disease. The treatment, now known as *therapeutic angiogenesis,* involves the injection of copies of the gene whose protein product directs the formation of new blood vessels around the aneurysm. The newly formed *collateral* arteries may be viewed as an auto-bypass (fig. 21.5).

There are several intriguing features of the therapy. Several years ago, researchers had discovered a protein growth factor, dubbed *vascular endothelial growth factor,* or VEGF, which promotes the sprouting of new blood vessels from pre-existing vessels. This is the phenomenon of angiogenesis, discussed earlier (chapter 7), in reference to cancerous growth. Indeed, VEGF was first discovered by cancer researchers, who noticed that tumor cells utilize VEGF to develop a microvessel network to facilitate delivery of nutrients and oxygen. Accelerated angiogenesis is also a normal event in fetal development. The subsequent isolation and cloning of the VEGF gene provided the impetus for Dr. Isner's study.

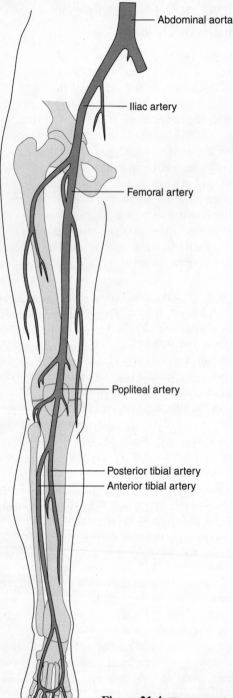

Abdominal aorta

Iliac artery

Femoral artery

Popliteal artery

Posterior tibial artery
Anterior tibial artery

Remarkably, the gene was transferred *intramuscularly* in the form of a *naked* plasmid DNA capable of encoding VEGF with its 165-amino acid sequence. In other words, the plasmid was not carried by a bacterium but was injected into smooth muscle cells adjacent to the blockage (fig. 21.5). Muscle cells are among the few in the body that are capable of receiving so-called naked DNA—that is, genes that are not packaged in bacteria or viruses. In several patients, direct intramuscular gene transfer effectively stimulated collateral vessel growth and provided relief for the limb ischemia (deficiency of blood flow).

With regard to correcting peripheral vascular disease, this is the first medical therapy that appears to be equivalent to successful surgical intervention. VEGF gene therapy holds promise for the nonsurgical correction of certain heart disorders. There may be one major drawback. The potentiation of local angiogenesis carries the *theoretical* risk that pathological tumor angiogenesis at a remote site could be stimulated.

THE FUTURE OF *HOMO SAPIENS*

The nature and circumstances of human life have been profoundly altered in the last few decades. Cultural innovations have occurred on an unprecedented scale. Cultural changes provide potential opportunities and potential disasters. Each innovative cultural change is rooted in new knowledge and is accompanied by new outlooks and responsibilities. However, too often new information is distorted to fit preexisting patterns of attitudes, beliefs, and actions. Despite the onrush of new knowledge, humans still remain encumbered by dogmas that trace their roots to early historic times. People are generally more comfortable in adhering to old values. However, many of our old values and prescribed actions are much too inflex-

Figure 21.4 The normal supply of arteries in the human leg. A blockage of the popliteal artery in the region of the knee would result in failure of blood flow in the lower extremities. The outcome is severe claudication, pain at rest, gangrene, and, if untreated, loss of the leg.

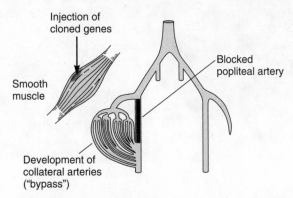

Figure 21.5 Direct intramuscular transfer of a cloned gene for a protein-growth factor (VEGF) promotes the formation of new blood vessels (collateral arteries) circumventing the blocked popliteal artery.

ible to cope adequately with the profound changes and variety of real-life situations that we are witnessing today. Adjustments in values have to be made when new knowledge in a changing situation does not reinforce past values, values based on a different set of circumstances.

Faced with the imperative need of establishing new values, how do we decide on new moral standards to guide our actions? The question of new values proceeds from the assumption that we know what we want. For the first time we have the capacity to engineer ourselves—to design ourselves and our future. We have intervened in nature and now have the capability of changing our own nature. Engineering the engineer, however, is quite different from engineering other organisms or mechanical objects. A particularly knotty

aspect of the problem is that the rapid advances in science and technology are known only in bits by a myriad of specialists in different branches. The individual in the street must now rely on specialists for information on nearly any problem that he or she encounters. More often than not, the specialists or authorities do not agree.

The last two decades have seen a surge of scientific knowledge and technological advances that few could have foreseen or dared to prophesy. Equally, we can be sure that the next two decades will contain surprises that promise an even greater strain on existing cultural and moral fabrics. This evinces either a tragic sense of despair that humans can become lost in their own machinations or a sober expectation that humans have the capabilities to manage constructively their own destiny.

EPILOGUE

"SCIENTIFIC CREATIONISM" IS NOT SCIENCE

Charles Darwin's explanation of evolution has been debated ever since the appearance of his book, *The Origin of Species.* The entire edition of 1,250 copies was sold out on the very day it appeared, November 24, 1859. Some press reviews were favorable; others were scathing or satiric. The most favorable comments of the work appeared in the London *Times,* comments by the distinguished scientist Thomas Huxley.

Darwin's theory appeared to many scientists to be refreshingly simple: some chance variations better adjust individuals to their environments, and such variant individuals tend to survive and transmit their favorable characteristics to descendants. This is the essence of natural selection, and it seemed harmless enough until Darwin hinted that humans might have evolved from "lower" forms of life. This notion was heretical because it contradicted the story of divine creation as told in the Bible. The biblical rendition is that plants and animals were created as separate species, all to be ruled over by humans, who were created themselves in the image of God. Religious opposition to Darwin's oblique suggestion that humans were not the unique crown of divine creation came from evangelists in Victorian England, who believed staunchly in the inerrancy of the Scriptures.

Today, the roots of opposition to Darwinism spring largely from Protestant fundamentalism, a sect that insists on the literal interpretation of the Bible. To preserve their absolutist concept of the Bible, the Protestant fundamentalists have woven Genesis into a spurious theory called "scientific creationism." The intent of scientific creationism is to give the story of Genesis a veneer of scientific respectability. The creationists have pressed for teaching their supernatural brand of evolution in the public schools. Scientific creationism is a mockery of the objectivity of science and debases conventional religion as well. Prominent theologians have been offended by the intrusion of religious belief into politics and public schools and by the contention that acts of faith such as divine creation can be construed as scientific. The enthralling Book of Genesis was never intended to be a textbook in science.

Scientific creationism is not a scientific alternative to Darwinian evolution. It demands the absolute acceptance of views not subject to test or revision. Scientific creationism extends beyond the simple suggestion that a Divine Being could have created the universe. Fundamentalist creationists would assert, for example, that the earth is less than 10,000 years old. This indefensible view is not even found in the mainstream of Christian theology. It is, however, a view that certain fundamentalist sects choose to read into the Book of Genesis. The fact that the earth is over four billion years old is not disputed by any serious scholar.

Scientific creationism is an inflexible view not open to empirical testing. Science necessarily rejects certainty and predicates acceptance of a concept on objective testing and the possibility of continual revision. The claims of the creationists are unverifiable (or unfalsifiable) and, hence, inherently unscientific. Scientific creationism is clearly the untestable belief of a particular religious sect. The creationists start with a conclusion, accept it as revealed truth not open to empirical testing, and then try to "prove" their contention—not by adducing positive evidence but by undermining evolutionary evidence.

The reaction of the public to the issues of scientific creationism has been largely one of confusion. The average person in the street has been bewildered by the complex arguments on both sides. Creationists challenge the validity of the data supporting evolutionary change, emphasizing the deficiencies in the fossil record. Evolutionists acknowledge that there are uncertainties in the details of evolutionary history; nevertheless, the body of evolutionary knowledge is continually growing and increasing in accuracy. The layperson is not generally conversant with the philosophy or methodology of science and lacks the confidence to weigh the issues. The local citizen perceives the controversy as a dispute between two groups of passionately committed individuals and supposes that one outlook is as good as another. When the creationists ask for equal time in the science classroom, the person in the street views the situation as a reasonable request for fair play. But the request is unreasonable because creationism only masquerades as science, and the integrity of science itself cannot be violated in the science classroom. Science does not resort to miraculous explanations. There may be good reasons to discuss theistic beliefs about creation in a school curriculum, but religious beliefs should be discussed in appropriate courses dealing with theology or comparative religion. The biblical account of creation, however disguised, should not be taught as if it were science—which it decidedly is not.

SUGGESTED READINGS

Agnew, N., and M. Demas. 1998. Preserving the Laetoli footprints. *Scientific American*, September, 44–55.

Alberts, B., D. Bray, L. Lewis, M. Raff, K. Roberts, and J. D. Watson. 1994. *Molecular Biology of the Cell.* 3d ed. New York: Garland Publishing.

Alters, B. J., and W. F. McComas. 1994. Punctuated equilibrium: The missing link in evolution education. *The American Biology Teacher* 56: 334–41.

Altman, L. K. 1993. Surprise discovery about split genes wins Nobel prize. *New York Times*, October 12, section C, p. 3.

Alvarez, W., and F. Asaro. 1990. What caused the mass extinction? An extraterrestrial impact. *Scientific American*, October, 78–84.

Andrews, P., and C. Stringer. 1989. *Human Evolution: An Illustrated Guide.* New York: Cambridge University Press.

Angier, N. 1990. Nature may fashion all cells' proteins from a few primordial parts. *New York Times*, December 11, section C, p.1.

———. 1994. Keys emerge to mystery of "junk" DNA. *New York Times*, June 28, section C, p.1.

Avise, J. C. 1994. *Molecular Markers, Natural History and Evolution.* London: Chapman and Hall.

Ayala, F.S. 1995. The myth of Eve: Molecular biology and human origins. *Science* 270: 1930–37.

Barlow, G. W. 1991. Nature-nurture and the debates surrounding ethology and sociobiology. *American Zoologist* 31: 286–97.

Beadle, G. W. 1980. The ancestry of corn. *Scientific American*, January, 112–19.

Berra, T.M. 1990. *Evolution and the Myth of Creationism: A Basic Guide to the Facts in the Evolution Debate.* Palo Alto, CA: Stanford University Press.

Bodmer, W. F., and L. L. Cavalli-Sforza. 1970. Intelligence and race. *Scientific American*, October, 19–29.

———. 1976. *Genetics, Evolution, and Man.* San Francisco: W. H. Freeman.

Bishop, J. A., and L. M. Cook. 1975. Moths, melanism and clean air. *Scientific American*, January, 90–93.

Brainard, J. 1998. Giving Neandertals their due: Similarities with modern humans shift the image of the caveman brute. *Science News* 154: 72–73.

Brown, G. D., Jr. 1995. *Human Evolution.* Dubuque, IA: WCB Publishers.

Cann, R. L., M. Stoneking, and A .C. Wilson. 1987. Mitochondrial DNA and human evolution. *Nature* 325: 31–36.

Capecchi, M. R. 1994. Targeted gene replacement. *Scientific American*, March, 52–59.

Cartmil, M. 1997. The third man. *Discover*, September, 56–61.

Cavalli-Sforza, L. I. 1991. Genes, peoples, and languages. *Scientific American*, November, 104–10.

Cavalli-Sforza, L. L. , P. Mennozzi, and A. Piazza. 1994. *The History and Geography of Human Genes.* Princeton, NJ: Princeton University Press.

Cherfas, J. 1991. Ancient DNA : Still busy after death. *Science* 253: 1354–56.

Clark, R. W. 1984. *The Survival of Charles Darwin*. New York: Random House.

Colburn, T., D. Dumanoski, and J. P. Myers. 1991. *Our Stolen Future—Are We Threatening Our Fertility and Survival—A Scientific Detective Story*. New York: Dutton Books.

Collins, F. S., M. S. Guyer, and A. Chakravati. 1997. Variations on a theme: Cataloging human DNA. *Science* 278: 1580–82.

Collins, F. S., A. Patrinos, E. Jordan, A. Chakravarti, R. Gesteland, and L. Walters. 1998. New goals for the U.S. human genome project: 1998–2003. *Science* 282: 682–89.

Committee on Science and Creationism. 1984. *Science and Creationism: A View from the National Academy of Science*. Washington, DC: National Academy Press.

Conroy, G. C. 1997. *Reconstructing Human Origins: A Modern Synthesis*. New York: W.W. Norton.

Coppens, Y. 1994. East side story: The origin of humankind. *Scientific American*, May, 88–95.

Courtillot, V.E. 1990. What caused the mass extinction? A volcanic eruption. *Scientific American*, October, 85–92.

Crick, F. H. C. 1998. *What Mad Pursuit*. New York: Basic Books. Crick's recollection of the epoch-making discovery.

Cullotta, E. 1995. Asian hominids grow older. *Science* 270: 1116–17.

Darwin, C. 1859. *On The Origin of Species by Means of Natural Selection*. London: John Murray. A fascimile of the first edition has been reprinted by Harvard University Press (1964).

———. 1960. *The Voyage of the Beagle (1840)*. New York: Bantam Books.

———. 1969. *The Autobiography of Charles Darwin, 1809–1882*. Edited with appendix and notes by his grand-daughter, Nora Barlow. New York: W.W. Norton.

Dawkins, R. 1986. *The Blind Watchmaker*. New York: W. W. Norton.

———. 1989. *The Selfish Gene*. 2d ed. London: Oxford University Press.

———. 1995. *River Out of Eden: A Darwinian View of Life*. New York: Basic Books.

———. 1995. God's utility function. *Scientific American*, November, 80–85.

———. 1996. *Climbing Mount Improbable*. New York: W.W. Norton.

deDuve, C. 1996. The birth of complex cells. *Scientific American*, April, 50–57.

DeRobertis, E. M., G. Oliver, and C. V. E. Wright. 1990. Homeobox genes in the vertebrate body. *Scientific American*, July, 46–52.

Dennison, R. 1993. Using Darwin's experimental work to teach the nature of science. *The American Biology Teacher* 55: 50–53.

Diamond, J. M. 1989. The cruel logic of our genes. *Discover*, November, 72–79.

———. 1991. Curse and blessing of the ghetto. *Discover*, March, 60–65.

———. 1993. What are men good for? *Natural History* 102: 24–29.

———. 1996. Why women change. *Discover*, July, 130–38.

———. 1997. Mr. Wallace's line. *Discover*, August, 76–83.

———. 1998. *Why is Sex Fun? The Evolution of Human Sexuality*. New York: Basic Books.

Dickerson, R.E. 1972. The structure and history of an ancient protein. *Scientific American*, April, 58–72.

Dobzhansky, T. 1951. *Genetics and the Origin of Species*, 3d ed. New York: Columbia University Press.

———. 1970. *Genetics and the Evolutionary Process*. New York: Columbia University Press.

———. 1973. Nothing in biology makes sense except in the light of evolution. *The American Biology Teacher* 35:125–29.

Drlica, K. 1992. *Understanding DNA and Gene Cloning: A Guide for the Curious*, 2d ed. New York: Wiley.

————. 1994. *Double-Edged Sword: The Promises and Risks of the Genetic Revolution*. Reading, MA: Addison-Wesley.

Dugatkin, L. A., and J. Godin. 1998. How females choose their mates? *Scientific American*, April, 56–61.

Dunn, L. C., and S. P. Dunn. 1957. The Jewish community of Rome. *Scientific American*, March, 118–28.

Eldredge, N. 1985. *Time Frames: The Rethinking of Darwinian Evolution and the Theory of Punctuated Equilibrium*. New York: Simon & Schuster.

Felgner, P. L. 1997. Nonviral strategies for gene therapy. *Scientific American*, June, 102–6.

Fischman, J. 1996. A fireplace in France. *Discover*, January, 69.

————. 1996. Evidence mounts for our African origins—and alternatives. *Science* 271: 1364.

Fischman, J., and T. Folger. 1996. Little foot, big implications. *Discover*, January, 68.

Fisher, R. A. 1958. *The Genetical Theory of Natural Selection*, 2d ed. New York: Dover.

Friedmann, J. 1997. Overcoming the obstacles for gene therapy. *Scientific American*, June, 96–101.

Futuyma, D. J. 1995. *Science on Trial: The Case for Evolution*. Sunderland, MA: Sinauer Associates.

Gerhart, J., and M. Kirschner. 1997. *Cells, Embryos, and Evolution*. Malden, MA: Blackwell Science.

Glass, B. 1953. The genetics of the Dunkers. *Scientific American*, August, 82–87.

Glausiusz, J. 1995. Hidden benefits. *Discover*, March, 30–31.

Goodfellow, P. 1995. A big book of the human genome. *Nature* 377: 285–86.

Gould, S.J. 1980. *The Panda's Thumb: More Reflections in Natural History*. New York: W.W. Norton.

————. 1987. Darwinism defined: The difference between fact and theory. *Discover*, January, 64–70.

————. 1989. *Wonderful Life: The Burgess Shale and the Nature of History*. New York: W.W. Norton.

————. 1991. *Bully for Brontosaurus*. New York: W.W. Norton.

————. 1995. *Dinosaur in a Haystack: Reflections in Natural History*. New York: Harmony Books.

————. 1996. *The Mismeasure of Man*. rev. ed. New York: W.W. Norton.

Grady, D. 1996. Tracing a genetic disease to bits of traveling DNA. *New York Times*, March 3, section C, p.1.

Grant, P. R. 1991. Natural selection and Darwin's finches. *Scientific American*, April, 82–87.

Grimaldi, D. A. 1996. Captured in amber. *Scientific American*, April, 84–91.

Hartl, D. L. 1999. *A Primer of Population Genetics*, 3d ed. Sunderland, MA: Sinauer Associates.

Hardy, G. H. 1908. Mendelian proportions in a mixed population. *Science* 28: 49–50.

Haseltine, W. A. 1997. Discovering genes for new medicines. *Scientific American*, March, 92–97.

Horgan, J. 1996. The world according to RNA. *Scientific American*, January, 27–30.

————. 1997. The sinister cosmos: A meteorite yields clues to life's molecular handedness. *Scientific American*, May, 18–21.

Hrdy, S. B., C. Janson, and C. van Schaik. 1995. Infanticides: Let's not throw the baby out with the bath water. *Evolutionary Anthropology* 3: 151-54.

Huckabee, C. J. 1989. Influences on Mendel. *The American Biology Teacher* 51: 84–89.

Huxley, A. 1932. *Brave New World*. New York: Harper & Row.

Huxley, J. 1953. *Evolution in Action*. New York: The New American Library.

Ingram, V. M. 1963. *The Hemoglobins in Genetics and Evolution*. New York: Columbia University Press.

Jameson, D. L., ed. 1977. *Evolutionary Genetics: Benchmark Papers in Genetics*, Vol. 8. Stroudsberg PA: Dowden, Hutchinson & Ross.

Johanson, D. C., and M. A. Edey. 1981. *Lucy: The Beginnings of Humankind*. New York: Simon & Schuster.

Johanson. D. C., and B. Edgar. 1996. *From Lucy to Language*. New York: Simon & Schuster.

Jukes, T. H. 1980. Silent nucleotide substitutions and the molecular evolutionary clock. *Science* 210: 973–78.

Katz-Sidlow, R. J. 1998. In the Darwin family tradition: Another look at Charles Darwin's ill health. *Journal of the Royal Society of Medicine* 91: 484–88.

Kiester E., Jr. 1991. A bug in the system. *Discover*, February, 70–76.

Kimura, M. 1979. The neutral theory of molecular evolution. *Scientific American*, November, 98–126.

Keller, E. F. 1983. *A Feeling for the Organism: The Life and Work of Barabara McClintock*. New York: W. H. Freeman.

Kettlewell, H. B. D. 1959. Darwin's missing evidence. *Scientific American*, March, 48–53.

Kitcher, P. 1982. *Abusing Science: The Case Against Creationism*. Cambridge: MIT Press.

Kreitman, M. 1996. The neutral theory is dead. Long live the neutral theory. *BioEssays* 18 (no. 8): 678–83.

Lack, D. 1947. *Darwin's Finches*. London: Cambridge University Press.

————. 1953. Darwin's finches. *Scientific American*, April, 66–72.

Lawn, R. M., A. Efstratiadis, C. O. O'Connell, and T. Maniatis. 1980. The nucleotiode sequence of the human *β*-globin gene. *Cell* 21: 644–51.

Leakey, M., and A. Walker. 1997. Early hominid fossils from Africa. *Scientific American*, June, 74–79.

Leakey, R. 1994. *The Origin of Humankind*. New York: Basic Books.

Leakey, R., and R. Lewin. 1992. *Origins Reconsidered: In Search of What Makes Us Human*. New York: Doubleday.

Levy, S. B. 1998. The challenge of antibiotic resistance. *Scientific American*, March, 46–53.

Lewin, R. 1994. *The Origin of Modern Humans*. New York: Scientific American Library.

————. 1996. Ancient humans found refuge in Java. *New Scientist* 152 (no.2061/2): 16.

Li, W., and D. Graur. 1991. *Fundamentals of Molecular Evolution*. Sunderland, MA: Sinauer Associates.

Lindahl, T. 1997. Facts and artifacts of ancient DNA. *Cell* 90: 1–3.

Linden, E., and M. Nichos. 1992. A curious kinship: Apes and humans. *National Geographic* 181: 32–46.

Lodish, H., D. Baltimore, A. Berk, S. D. L. Zipursky, P. Matsudaira, and J. Darnell. 1995. *Molecular Cell Biology*. 3d ed. New York: Scientific American Books.

Madigan, M. T., and B. L. Marrs. 1997. Extremophils. *Scientific American*, April, 82–87.

Mange, E. J., and A. P. Mange. 1998. *Basic Human Genetics*. 2d ed. Sunderland, MA: Sinauer Associates.

Margulis, L., 1981. *Symbiosis in Cell Evolution*. San Francisco: W.H. Freeman.

Margulis, L., and D. Sagan. 1986. *Microcosms: Four Billion Years of Evolution from Our Microbial Ancestors*. New York: Simon & Schuster.

————. 1986. *Origin of Sex: Three Billion Years of Genetic Recombination*. New Haven: Yale Unviersity Press.

Margulis, L., and K.V. Schwartz. 1987. *Five Kingdoms: An Ilustrated Guide to the Phyla of Life on Earth*. San Francisco: W.H. Freeman.

Marks, J. 1995. *Human Biodiversity: Genes, Races, and History*. New York: Aldine de Gruyter.

May, R. M. 1992. How many species inhabit the earth? *Scientific American*, October, 42–48.

Maynard Smith, J. 1978. *The Evolution of Sex*. Cambridge: Cambridge University Press.

Mayr, E. 1991. *One Long Argument: Charles Darwin and the Genesis of Modern Evolutionary Thought*. Cambridge, MA: Harvard University Press.

McComas, W. F., and B. J. Alters. 1994. Modeling modes of evolution: Comparing phyletic gradualism and punctuated equilibrium. *The American Biology Teacher* 56 (no. 6): 354–61.

McConkey, E. H. 1993. *Human Genetics: The Molecular Revolution*. Boston: Jones and Bartlett.

McGinnis, W., and M. Kuziora. 1994. The molecular architects of body design. *Scientific American*, February, 58–66.

McInerney, J. D. 1995. The human genome project and biology education. *BioScience* 45 (no. 11): 786–92.

McInerney, J. D., and R. Moore. 1993. Voting in science: Raise your hand if you want humans to have 48 chromosomes. *The American Biology Teacher* 55 (no. 3): 132–34.

McKinney, M. L. 1998. The juvenilized ape myth—our "overdeveloped" brain. *Bioscience* 48 (no. 2): 109–17.

McKusick, V.A. 1997. *Mendelian Inheritance in Man: A Catalog of Human Genes and Genetic Disorders.* 12th ed. Baltimore: Johns Hopkins University Press. (Internet: http://www3.ncbi.nlm.nih.gov/omim/)

Mendel, G. 1866. "Versuche über pflanzenhybriden." *Verhandlungen des Naturforschenden Vereines in Brünn* 4: 3–47. English translations appear in Peters 1969, Stern and Sherwood 1966, and Jameson 1977. (Internet: http//www.netspace.org/ MendelWeb)

————. 1986. Experiments in a Monastery Garden. *American Zoologist* 26: 749–52. Richard M. Eakin performs on stage as Mendel before a modern assemblage of biologists.

Menon, S. 1997. Enigmatic apes, *Discover,* August, 26–27.

Miller, S. L. 1953. A production of amino acids under possible primitive earth conditions. *Science* 117: 528–29.

Milner, R. 1995. Charles Darwin: The last portrait. *Scientific American,* November, 78–79.

————. 1996. Keeping up Down House. *Natural History* 105: 54–60.

Monod, J. 1971. *Chance and Necessity.* New York: Knopf.

Montagu, A. 1964. *The Concept of Race.* New York: Free Press of Glencoe.

————. 1965. *Man's Most Dangerous Myth: The Fallacy of Race.* New York: Macmillan.

Moore, J. A. 1993. *Science as a Way of Knowing: The Foundations of Modern Biology.* Cambridge: Harvard University Press.

Moore, R. 1997. The business of creation. *The American Biology Teacher* 59: 196–97.

————. 1997. The persuasive Mr. Darwin. *BioScience:* 47 (no. 2): 107–13.

————. 1998. Creationism in the United States I. Banning evolution from the classroom. *American Biology Teacher* 60: 486–507.

————. 1998. Creationism in the United States II. The aftermath of the Scopes Trial. *American Biology Teacher* 60: 568–77.

Morris, D. 1967. *The Naked Ape.* New York: Dell.

Mulligan, R. 1993. The basic science of gene therapy. *Science* 260: 926–32.

Mullis, K. B. 1990. The unusual origin of the polymerase chain reaction. *Scientific American,* April, 56–65.

Neel, J. V. 1994. *Physician to the Gene Pool: Genetic Lessons and Other Stories.* New York: Wiley.

————. 1995. New approaches to evaluating the genetic effects of the atomic bombs. *American Journal of Human Genetics* 57: 1263–66.

Nei, M. 1987. *Molecular Evolutionary Genetics.* New York: Columbia University Press.

Nemecek, S. 1997. Amphibians on-line. *Scientific American,* March, 18.

Nesse, R M., and G. C. Williams. 1998. Evolution and the origins of disease. *Scientific American,* November, 86–93.

Novick, G. E., C. C. Novick, J. Yunis, E. Yunis, P. A. DeMayolo, W. D. Cheer, P. L. Deininge, M. Stoneking, D. S. York, M. A. Batzer, and R. J. Herrera. 1998. Polymorphic Alu insertions and the Asian origin of native American populations. *Human Biology* 70 (no. 1): 23–40.

Ohta, T. 1996. The current significance and standing of neutral and nearly neutral mutation theories. *BioEssays* 18 (no. 8): 673–77.

Oliwenstein, L. 1997. How salt can kill. *Discover,* January, 90.

Oparin, A. I. 1953. *The Origin of Life.* New York: Dover.

Orgel, L.E. 1998. The origins of life—a status report. *The American Biology Teacher* 60 (no. 1): 10–13.

Pääbo, S. 1993. Ancient DNA. *Scientific American,* May, 86–92.

Padian, K., and L. M. Chiappe. 1998. The origin of birds and their flight. *Scientific American*, February, 38–47.

Poulton, J. 1987. All about Eve. *New Scientist*, 14 May, 51–53.

Pielou, E. C. 1991. *After the Ice Ages: The Return of Life to Glaciated North America*. Chicago: University of Chicago Press.

Penny, D., and C. J. O'Kelly. 1991. Seeds of a universal tree. *Nature* 6314: 106–8.

Peters, J. A., ed. 1959. *Classic Papers in Genetics*. Englewood Cliffs, NJ: Prentice-Hall.

Radash, L. 1989. The age-of-the-earth debate. *Scientific American*, August, 90–96.

Raff, R. A., and T. C. Kaufman. 1983. *Embryos, Genes, and Evolution*. New York: Macmillan

Reed, T. E. 1969. Caucasian genes in American Negroes. *Science* 165: 762–68.

Relethford, J. H. 1997. *The Human Species: An Introduction to Biological Anthropology*, 3d ed. Mountain View, CA: Mayfield Publishing.

Richardson, S. 1994. Mergers and acquisitions. *Discover*, December, 35.

Ridley, M. 1996. *Evolution*, 2d ed. London: Blackwell Science.

Rimoin. D. L., J. M. Conner, and R. E. Pyeritz. 1997. *Principles and Practice of Medical Genetics*, 3d ed. New York: Churchill Livingston.

Ross, P .E. 1990. All about Eve. *Scientific American*, April, 22D–24.

————. 1991. Hard words. *Scientific American*, April, 138–47.

————. 1992. Eloquent remains. *Scientific American*, May, 115–25.

Sagan, C. 1977. *The Dragons of Eden: Speculations on the Evolution of Human Intelligence*. New York: Ballantine Books.

Sapienza, C. 1990. Parental imprinting of genes. *Scientific American*, October, 52–61.

Sarich, V. M., and A. C. Wilson. 1966. Quantitative immunochemistry and the evolution of primate albumins. *Science* 154: 1563–66.

Sayre, A. 1975. *Rosalind Franklin and DNA*. New York: Norton.

Schluter, D. 1996. Ecological causes of adaptive radiation. *American Naturalist*: 148: 511–36.

————. 1996. Adaptive radiation along genetic lines of least resistance. *Evolution* 50: 1766–75.

Schull, W. J., M. Otake, and J. V. Neel. 1981. Genetic effects of the atomic bombs: A reappraisal. *Science* 213: 220–28.

Scriver, C. R., A. L. Beaudet, W. S. Sly, and D. Valle. eds. 1995. *The Metabolic and Molecular Basis of Inherited Disease*. 7th ed. New York: McGraw-Hill. (Also available in a CD-Rom version.)

Selander, R. K., A. G. Clark, and T. S. Whittam. 1991. *Evolution at the Molecular Level*. Sunderland, MA: Sinauer Associates.

Shipman, P. 1991. Hotheads. *Discover*, April, 19.

Shreeve, J. 1990. Argument over a woman. *Discover*, August, 52–59.

Sibley, C. G., and J. E. Ahlquist. 1990. *Phylogeny and the Classification of Birds: A Study in Molecular Evolution*. New Haven: Yale University Press.

Simpson, G. G. 1951. *Horses*. New York: Oxford University Press.

————. 1953. *Life of the Past*. New Haven: Yale University Press.

————. 1967. *The Meaning of Evolution*. New Haven: Yale University Press.

————. 1983. *Fossils and the History of Life*. New York: Scientific American Library.

————. 1994. *Tempo and Mode of Evolution*. New York: Columbia University Press.

Smith, J. M. 1990. The Y of human relationships. *Nature* 344 (no. 6267): 591–93.

Sober, E., and D. S. Wilson. 1998. *Unto Others: The Evolution and Psychology of Unselfish Behavior*. Cambridge: Harvard University Press.

Spiess, E.B. 1989. *Genes in Populations*. 2d ed. New York: John Wiley.

Stanley, S. M. 1998. *Children of the Ice Age: How Global Catastrophe Allowed Humans to Evolve*. San Francisco: W. H. Freeman.

Stebbins, G. L. 1950. *Plant Variation and Evolution*. New York: Columbia University Press.

————. 1982. *Darwin to DNA, Molecules to Humanity*. San Francisco: W. H. Freeman.

Stebbins, G. L., and F. J. Ayala. 1985. The evolution of Darwinism. *Scientific American*, July, 72–82.

Stern, C. 1943. The Hardy-Weinberg Law. *Science* 97: 137–38.

————. 1949. *Principles of Human Genetics*. San Francisco: W. H. Freeman.

Stern, C., and E. Sherwood. 1966. *The Origins of Genetics: A Mendel Source Book*. San Francisco: W. H. Freeman.

Strickberger, M.W. 1996. *Evolution*. 2d ed. Boston: Jones and Bartlett.

Stringer, C. B. 1990. The emergence of modern humans. *Scientific American*, December, 98–104.

Stringer, C. B., and C. Gamble. 1993. *In Search of the Neanderthals: Solving the Puzzle of Human Origins*. New York: Thames and Hudson.

Svitil, K. A. 1996. We are all Panamanians. *Discover*, April, 30–31.

Tattersall, I. 1995. *The Fossil Trail: How We Know What We Know About Human Evolution*. New York: Oxford University Press.

————. 1997. Out of Africa again . . . and again? *Scientific American*, April, 60–67.

Templeton, A. R. 1992. Human origins and analysis of mitochondrial DNA sequences. *Science* 255: 737.

Thorne, A. G., and M. H. Wolpoff. 1992. The multiregional evolution of humans. *Scientific American*, April, 76–83.

Thurman, E. 1993. *Human Chromosomes: Structure, Behavior and Effects*. New York: Springer-Verlag.

Tourney, C. P. 1994. God's own scientists. *Natural History* 103 (no. 7): 4–9.

Verma, I. M. 1990. Gene therapy. *Scientific American*, November, 68–84.

Wallace, A. R. 1908. *My Life: A Record of Events and Opinions*. rev. ed. London: Chapman and Hall.

Wallace, D. C. 1997. Mitochondrial DNA in aging and disease. *Scientific American*, August, 40–47.

Wallace, D. C., and C. Torroni. 1992. American Indian prehistory as written in the mitochondrial DNA: A review. *Human Biology* 64: 403–17.

Watson, J. D. 1968. *The Double Helix*. New York: Atheneum. Watson's forthright account of the race to discover the structure of DNA.

Watson, J. D., and F. H. C. Crick. 1953. A structure for deoxyribose nucleic acid. *Nature* 171: 737. The original scientific paper (one page!) on the announcement of the chemical architecture of DNA.

Watson, J. D., N. H. Hopkins, J.W. Roberts, J. A. Stertz, and A. M. Weiner. 1987. *Molecular Biology of the Gene*. 4th ed. Menlo Park: Benjamin Cummings.

Weiner, J. 1994. *The Beak of the Finch: Evolution in Real Time*. New York: Alfred A. Knopf.

Welsh, M. J., and A. E. I. Smith. 1995. Cystic fibrosis. *Scientific American*, December, 52–60.

White, R., and J. M. Lalouel. 1988. Chromosome mapping with DNA markers. *Scientific American*, February, 40–48

White, T. D., G. Suwa, and B. Asfaw. 1994. *Australopithecus ramidus*, a new species of early hominid from Aramis, Ethiopia. *Nature* 371: 306–12.

Whittaker, R. H. 1969. New concepts of kingdoms of organisms. Science 163:150–60.

Whittled, P. 1993. *From So Simple A Beginning: The Book of Evolution*. New York: Macmillan.

Wills, C. 1995. When did Eve live? An evolutionary detective story. *Evolution*: 49: 593–608.

Wilson, A. C. 1985. The molecular basis of evolution. *Scientific American*, October, 164–73.

Wilson, A. C., and R. L. Cann. 1992. The recent African genesis of humans. *Scientific American*, April, 68–73.

Wilson, A. C., and V. M. Sarich. 1969. A molecular time scale for human evolution. *Proceedings for the National Academy of Sciences* 63: 1088–93.

Wilson, E. O. 1975. *Sociobiology: The New Synthesis*. Cambridge: Harvard University Press.

————. 1978. *On Human Nature*. Cambridge: Harvard University Press.

————. 1984. *Biophilia: The Human Bond with Other Species*. Cambridge: Harvard University Press.

————. 1992. *Diversity of Life*. Cambridge: Harvard University Press.

————. 1998. *Consilience: The Unity of Knowledge*. New York: Alfred A. Knopf.

Wolpoff, M. H. 1998. Neandertals: Not so fast. *Science* 282: 1991.

Woolf, C. M., and F. C. Dukepoo. 1969. Hopi Indians, inbreeding, and albinism. *Science* 164: 30–37.

Wright, S. 1968. *Evolution and The Genetics of Populations*. 4 vols. Chicago: University of Chicago Press.

Zimmer, C. 1995. Coming onto the land. *Discover*, June, 119–27.

INDEX

A